How to be Innovative

Early Stage Innovation for Scientists, Technologists and Others — From Idea to Proof-of-Concept

How to be Innovative

Early Stage Innovation for Scientists, Technologists and Others — From Idea to Proof-of-Concept

Peter W Lednor
Peter Lednor Consulting, The Netherlands

World Scientific

NEW JERSEY • LONDON • SINGAPORE • BEIJING • SHANGHAI • HONG KONG • TAIPEI • CHENNAI • TOKYO

Published by

World Scientific Publishing Co. Pte. Ltd.
5 Toh Tuck Link, Singapore 596224
USA office: 27 Warren Street, Suite 401-402, Hackensack, NJ 07601
UK office: 57 Shelton Street, Covent Garden, London WC2H 9HE

British Library Cataloguing-in-Publication Data
A catalogue record for this book is available from the British Library.

HOW TO BE INNOVATIVE
Early-Stage Innovation for Scientists, Technologists and Others —
From Idea to Proof-of-Concept

ISBN 978-981-3222-02-1
ISBN 978-981-3222-03-8 (pbk)

For any available supplementary material, please visit
https://www.worldscientific.com/worldscibooks/10.1142/10491#t=suppl

"Every organization – not just businesses – needs one core competence: innovation."

P. Drucker, Management Issues for the 21st Century, 2001

"Innovation is inefficient. More often than not, it is undisciplined, contrarian, and iconoclastic; and it nourishes itself with confusion and contradiction. In short, being innovative flies in the face of what almost all parents want for their children, most CEOs want for their companies, and heads of states want for their countries. And innovative people are a pain in the ass."

N. Negroponte, "Creating a culture of ideas",
MIT Technology Review, Feb. 2003

"And it ought to be remembered that there is nothing more difficult to take in hand, more perilous to conduct, or more uncertain in its success, than to take the lead in the introduction of a new order of things. The innovator has for enemies all those who have done well under the old conditions and lukewarm defenders in those who may do well under the new."

Machiavelli, The Prince, 1505

Preface

The topic of innovation has received much attention in recent years. Companies innovate to remain competitive. Governments, national or local, wish to stimulate and support innovation, and look to innovation to provide new or improved services for their citizens. Universities often aim to commercialize, at least in part, new knowledge that they generate.

In addition to the perspectives from this wide variety of organizations, there are many facets to innovation: the relation to organizational strategy, the definition of the scope for innovation, the generation, development and handover of ideas, open innovation, and issues of trust, communication and learning. This breadth of innovation, in the organizations involved as well as in the topics represented, has generated many books and articles. As a result, it can be difficult for those starting an involvement with innovation, as an innovator, a customer of innovation, or a manager of innovation, to quickly obtain an understanding of the topic.

This book aims to meet this need, by providing a concise guide to key features of innovation, illustrated with practical examples, and supported by extensive references which allow further exploration of the many topics covered. There are three areas of focus.

Firstly, the book concentrates on early stage innovation: from defining desirable areas for innovation through idea generation and development until a Proof-of-Concept has been established. At this point sufficient confidence in the idea has been established to make implementation a serious goal. This first stage of innovation, up to Proof-of-Concept, has features which are common across many areas of innovation. The second main stage of innovation, from Proof-of-Concept to a successful launch, requires attention to be increasingly paid to the specific market addressed.

A second area of focus is on the needs of scientists and technologists. Much writing on innovation is from business, organizational or strategic perspectives, which are less familiar to scientists and technologists, particularly in the early stages of their careers.

Thirdly, the book is focused on innovation by organizations, rather than by individual entrepreneurs. Many people join an organization with a commitment to innovation; indeed, as the prefatory quote by Peter Drucker makes clear, this should be all organizations.

The book provides a set of resources — concepts, practical examples, and literature — which will help individuals, and their organizations, to become more effective in being innovative. Numerous organizations, particularly large ones, will already have some approach to innovation in place; the ideas and examples presented here may clarify gaps in understanding or practice, and suggest ways to address these. Small companies often rely on quick, informal decision-making but growth will sooner or later require the development of processes and policies. The content of this book will help prepare for this future before it arrives.

The book is structured as follows. Chapter 1 deals with key concepts in innovation. A shared understanding of what innovation means within an organization is important to achieving it. Chapter 2 deals with systematic processes for defining areas for innovation and moving through to establishment of Proof-of-Concept. While some may argue that innovation requires maximum freedom (and funding) untrammeled by the needs of formal review processes, it is clear that innovation benefits from a systematic approach, and should not be left to chance; this is even truer in the case of the multidisciplinary, geographically-dispersed teams which are becoming increasingly important. Chapter 3 deals with open innovation, which refers to the mechanisms by which an organization interacts with the outside world, bringing in, or spinning out, knowledge and technologies. Chapter 4 is concerned with the "softer" issues of behaviors, teams, communication and learning, which are key components of efforts to be innovative, and which are topics whose importance becomes clearer as experience with innovation develops. Chapter 5 consists of eight case studies covering a range of organizations, topics and countries, focusing on examples at the level of projects. Illustrating early stage innovation at the level of projects is likely to be more useful, in the context of this book, than higher level analyses such as comparison of companies. In addition to providing illustration of the material in the preceding chapters, this chapter offers the reader the opportunity to learn from other organizations and projects. Chapter 6 discusses some myths and pitfalls, further steps the reader can take, and some likely developments in the field of innovation management.

The topic of (early stage) innovation will undoubtedly continue to develop in coming years, both in terms of understanding and through examples in practice. This book provides a context and foundation for such developments.

Peter Lednor
Heemstede, The Netherlands
December, 2017

About the Author

Peter Lednor combines a technical career in industrial Research and Development with extensive experience of managing, and teaching, innovation. He holds a B.Sc. degree (University of Manchester, 1971) and a Ph.D. (University of Sussex, 1974), both in chemistry. Following post-doctoral fellowships at the University of Munich, and at CNRS, Paris, France, he joined the Shell Technical Center in Amsterdam in 1978, where he worked until retirement in 2009. He is the author of some 40 academic publications and 28 patents.

His R&D career at Shell included exploratory research, development projects, and innovation management. Initiating and managing external relationships with companies and universities played an important role throughout his career. He was co-chair of four international conferences on advanced catalytic materials, organized through the Materials Research Society, and was the industrial supervisor of 4 Ph.D. students. He was an Associate Fellow of Saïd Business School, University of Oxford from 2007 to 2015. His work there covered program director roles in Executive Education as well as lecturing.

In May 2010 he founded Peter Lednor Consulting, focusing on innovation. He was awarded a fellowship by the São Paulo Research Foundation (Fapesp) to spend six weeks in 2010 as a visiting professor at the University of Campinas, Brazil, where he taught a post-graduate course on innovation, and worked on technology transfer issues with Inova, the technology transfer office at University of Campinas. Since then he has provided courses or lectures on innovation in the Netherlands, UK, France, Brazil, USA, India and Kazakhstan, for a variety of organizations.

Acknowledgements

This book has been developed from courses given in the period 2010–17, for a variety of organizations in a number of different countries. Participants provided questions and comments which helped shape the content of the book, and I am grateful for the many positive and encouraging responses. I would particularly like to thank Prof. Telma Franco for inviting me to spend six weeks in 2010 at the University of Campinas, Brazil, where the first of these courses was given, and to Prof. Roberto Lotufo, who invited me to spend part of my time there in Inova, the technology transfer office at the University of Campinas.

I am very grateful to the authors of the case studies in Chapter 5 for the time and effort involved in producing their contributions. Examples of how innovation happens, in different areas and in different contexts, provide valuable illustrations of the material discussed in other chapters.

From 2001–2007 I worked in various "GameChanger" teams in the Shell technology organization; these teams set out to develop a culture of innovation, involving new ways of working as well as providing funding for new projects. It was a period characterized by considerable enthusiasm for, and learning about, innovation. I would like thank Dave Austgen and Joe Machado for the opportunities they provided, as well as the many other colleagues, too numerous to mention individually, whose contributions were stimulating, challenging or entertaining.

I would like to thank Prof. Rafael Ramirez and Dr. Marc Thompson at Saïd Business School, University of Oxford, for discussions over several years which broadened my perspectives on innovation.

I am grateful to the following reviewers who provided helpful comments: Shaun Lednor (Chapter 1), an anonymous referee (Chapter 2), Herman van Wechem (Chapter 3), and Richard Chorley (Chapter 4). Any errors or omissions are my responsibility.

And last but not least, heartfelt thanks to my family, who provided welcome distraction from, as well as support for, the tasks of authorship.

Contents

List of Figures

List of Tables

1 Innovation: What Is It, Why Do It?

1.1 INTRODUCTION

The topic of innovation is rarely out of the news. Many organizations — companies, governments, universities and others — proclaim the importance of being innovative, and the word innovation turns up in countless books, articles, TV and radio programs, and conversations. Despite all this attention, or perhaps because of it, there is often confusion about what innovation is, why it is important, and how best to ensure that efforts to become innovative are successful. This is a problem for people without experience of innovation who are called upon to be innovative. Such requests or expectations take many forms, for example, in writing (research) proposals, in establishing collaborations, and in developing new projects. This book provides a set of resources — concepts, practical examples, and literature references — which will help those who wish to become (more) innovative. This covers the output — the innovations — as well as the skills and organizational capacity which produce the output.

Very few innovations can be completely developed by a single individual, and hence innovation requires collaborations. These can be in the form of an organization in which the various skills required to make the innovative idea a success are present, or in the form of a network of partners developed by an individual or a team. In both cases, a common understanding of what innovation implies is necessary: there may be many different views on what innovation means, and it may be incorrectly assumed that the use of the same word implies the same understanding.

So what is innovation? In essence innovation deals with bringing a new idea successfully to market, but this apparently simple description raises questions about what new means, and to whom, what the market refers to, and what successful means. These points, as well as definitions of innovation, are discussed below, but it is useful to first survey the breadth of innovation by examining some of the organizations, and topics, in which innovation plays a role. Many people will think of new physical products when asked for examples of innovation, but innovation is also significant in providing services to meet social needs, for example distance learning, and alleviation of poverty, and in tackling major issues such as climate change, and ensuring sufficient, safe and affordable access to health care, energy, water and food.

1.2 THE BREADTH OF INNOVATION

This breadth of innovation is illustrated by the following examples, and is also reflected in the quote at the start of this book: "Every organization, not just business, needs one core competence: innovation" (Drucker, 2001). Core competencies are the combination of resources and skills which provide the foundation of companies' competitiveness.

Companies

It is essential for companies to innovate, for the simple reason that if they do not, they will be overtaken by a company which does. New and improved products involve innovation, in the features offered which meet customer needs, or through cost reductions leading to a lower price. Statements by company executives on the importance of innovation to their business are very common. Typical commercial innovations cover:

New products

Digital devices will frequently come to mind when thinking about innovative products, but a glance around shows that many, if not all, features of modern life represent innovations: heating, lighting, cooling (both of food and as air conditioning), transport, reading glasses, computers, radio and television are just a few examples of successful introductions of new ideas that have had a major impact on the world.

New manufacturing processes

For many centuries goods were manufactured by hand, with one person, or a small group, producing a complete product. The introduction of the assembly line, in which workers repeat a manufacturing step on parts which pass by on a moving belt, was a significant innovation in manufacturing technology. It is usually associated with the Ford Motor Company through its use, starting in 1913, for the manufacture of the Model T Ford. However, the history of the assembly line goes back much further: the Venetian Arsenal, which dates back to about 1104, assembled ships which passed down a canal and were fitted out by different shops along the way.

In the present period, 3D printing represents a new innovation in manufacturing. The technology is undergoing extensive development as a way of creating products by building them up layer by layer: this is sometimes referred to as additive manufacturing, in contrast to subtractive manufacturing in which a component is made by removing material from a larger piece, for example in wood or metal working.

New services

The growth of services relative to manufacturing has been a significant development in industrialized countries, and covers various sectors, including financial services, information technology and tourism. In addition, there are moves towards products being leased as part of a service, rather than being sold; examples include computers (IBM) and jet engines (GE). This trend represents innovations in business models and in customer relationships. The study and understanding of innovation in services is relatively recent, and also has to take into account the rapid development and application of digital platforms which form the basis of social media, as well as of companies such as Uber or Airbnb.

Governments

A common view of innovation is that it relies on individuals with a good idea, which is then developed using their own financial resources or attracting investors, ultimately becoming a commercial success. In this view there is little recognition of a role for governments. However, awareness of the role that governments play in stimulating innovation is growing. In reviewing government policy with respect to innovation, Patanakul and Pinto write "innovation continues to serve as a fundamental driver of the economic health of national economies particularly when it is actively and effectively supported by governmental agencies" (Patanakul and Pinto, 2014). The importance of innovation was stressed by President Obama in his 2011 State of the Union speech in which he devoted significant time to the topic of innovation, saying "The first step in winning the future is encouraging American innovation," and going on to say: "Throughout our history, our government has provided cutting-edge scientists and inventors with the support that they need. That's what planted the seeds for the Internet. That's what helped make possible things like computer chips and GPS. Just think of all the good jobs — from manufacturing to retail — that have come from these breakthroughs" (Obama, 2011).

The role of government, in the US, in stimulating innovation has been examined in a recent book entitled, *The Entrepreneurial State*. In contrast to the widespread perception that US innovation results from free markets and small government, the author makes the case that "from the Internet to biotech and even shale gas, the US State has been the key driver of innovation-led growth — willing to invest in the most uncertain phase of the innovation cycle and let business hop on for the easier ride down the way" (Mazzucato, 2014).

Within Europe, The Innovation Union "is the European Union strategy to create an innovation-friendly environment that makes it easier for great ideas to be turned into products and services that will bring our economy growth and jobs." This initiative is supported by Horizon 2020, an EU Research and Innovation program with nearly €80 billion of funding available over the seven years from 2014 to 2020 (EU Innovation Union, 2017).

Local government is also active in innovation. Many cities now have a chief innovation officer. European cities can compete for an award, made every two years, for the title of "Innovation Capital" (European Capital of Innovation award, 2017). The first prize is one million Euros, and the award criteria state that cities have to prove how they created local conditions to innovate by:

- **Experimenting** innovative concepts, processes, tools, and governance models as a test-bed for innovation

- **Engaging** citizens in the innovation process and ensuring the uptake of their ideas
- **Expanding** the city's attractiveness to become a role model for other cities
- **Empowering** the local ecosystem through the implementation of innovative practices.

The first award, in 2014, was given to Barcelona and the second to Amsterdam in 2016.

Many governments, at a national or local level, are also focused on stimulating innovation through creating or supporting clusters of companies. Silicon Valley, Boston's Route 128 corridor, and North Carolina's Research Triangle Park are well known examples; more recent ones include East London Tech City, New York Digital City and Start-up City Tel Aviv.

Further examples could be given, but these demonstrate the growing importance of innovation to national and local governments.

Universities

Universities have several interfaces with innovation: in teaching the subject as part of undergraduate or executive education programs, and levels in between such as Masters or PhD programs; in undertaking research into innovation in order to understand it as an economic and social phenomenon, and by commercialization of research findings through the creation of start-ups, or through technology transfer offices. The role of universities in collaborations and technology transfer is discussed further in Chapter 3.

Social innovation

Another major area of innovation, less well-known than business and technology innovation, is that of social innovation. This has been defined as "A novel solution to a social problem that is more effective, efficient, sustainable, or just than existing solutions and for which the value created accrues primarily to society as a whole rather than private individuals" (Phills, Deiglmeier and Miller, 2008). Examples include the Open University in the UK, and other models for distance learning, microfinance (for which the Grameen Bank and Muhammad Yunus shared the 2006 Nobel Peace Prize), and organizations such as Fair Trade, Greenpeace and Oxfam (Mulgan, 2007).

The process by which social innovations occur has been described, and compared to business and technology innovations (Mulgan, 2011). In all cases the stages involve an understanding of needs, the importance of trial and error, demonstrating a proof-of-principle in the form of a pilot or prototype, and dealing with the challenge of scale-up. This last phase seems to be more difficult in social innovations, where public acceptance, and willingness of governments to change, can be a slow process; in addition business can be more effective in replacing managers when necessary.

Societal challenges

Major challenges facing the world such as climate change, and provision of health care, food and water require innovative solutions. These involve technical innovations, but also require new business models, new partnerships and engagement with policy or regulatory issues. Technical advances behind the falling costs and rising performance of solar cells have been matched with innovative business models involving feed-in tariffs, through which individual households can export surplus electricity back to a grid. Health care anticipates major benefits from personalized medicine and telemedicine but at the same time must deal with issues around privacy and data security. Providing drinkable and affordable water requires (new) purification technologies, as well as decisions on water rights and managing water consumption. Provision of food on a global basis is affected by decisions on genetically modified crops or use of pesticides, which are influenced by public perceptions and government regulations. Improving urban mobility can include policy measures such as congestion charges, tax incentives for electric vehicles, new technologies such as batteries or fast charging points, and social schemes such as vehicle sharing, in turn supported by innovations in IT.

It is clear that these societal challenges require innovations at technical, business, governmental and social levels. A shared understanding between these sectors of what innovation means, processes to achieve it, and how to overcome barriers will help in finding effective solutions.

1.3 THE SIGNIFICANCE OF THE BREADTH OF INNOVATION

Awareness of this breadth of innovation is important for several reasons: it explains the level of interest in innovation, it accounts for the confusion which sometimes arises, it provides an opportunity to learn from other areas, and it can contribute to providing new solutions to problems.

Level of interest

The above brief survey gives some idea of the many areas in which innovation is found, and the wide range of bodies involved. In addition, the topic of innovation has many aspects, including creativity, idea development, business development, resource allocation, partnerships, culture and teamwork. Each of these subjects is represented by an extensive literature and many experts. As a result, the amount of information available, and the range of topics involved, can be discouraging, and sometimes confusing, for those wanting to explore the topic.

The literature on innovation is huge. An Internet search (Google) on the terms "innovation," "innovation definition" and "innovation management" in March 2017 returned 574 million, 124 million and 18.5 million hits, respectively. It can also be hard to know where to start when faced with articles variously entitled "The 5 myths of innovation" (Birkinshaw, Bouquet and Barsoux, 2011); "The six keys to creating new-growth businesses" (Christensen, Raynor and Anthony, 2003); "The seven myths of innovation" (Sawhney and Wolcott, 2004); "The eight essentials of innovation" (de Jong, Marston and Roth, 2015); "The 9 rules of innovation" (Satell, 2016); "The 10 types of innovation" (Keeley, Pikkel, Quinn and Walters, 2013); "The 12 different ways for companies to innovate" (Sawhney, Wolcott and Arroniz, 2006); as well as at least 25 definitions of innovation, some of which are contradictory (Anthony, 2012).

Confusion

Confusion arises about what an innovation is, as distinct from a new idea or an invention. For example, in a book on the sources of good ideas

(Johnson, 2010) the author writes "The academic literature on innovation and creativity is rich with subtle distinctions between innovations and inventions.....I have deliberately chosen the broadest possible phrasing — good ideas — to suggest the cross-disciplinary vantage point I am trying to occupy." This point is reflected in the title of the book, *Where Good Ideas Come From: The Seven Patterns of Innovation*, as well as in the "Chronology of Key Innovations 1400–2000" which includes global warming, the observation of gamma ray bursts from outer space, and evidence for an asteroid impact on the earth which led to the extinction of the dinosaurs. These are undoubtedly major ideas, based on evidence, but it is hard to argue that they meet the definition of innovation, given below, of "the successful exploitation of a new idea." The distinction between new ideas and innovation is important, since those aiming to be innovative will not, or at least should not, be satisfied with just the production of new ideas. Successful exploitation of a new idea is what defines an innovation.

The belief "I was innovative — my PhD thesis was full of new ideas" was regularly expressed in one form or another by participants in courses on which this book is based. This belief also confuses new ideas and innovation, and additionally adds research into the mix. The links between research and innovation are discussed later in this chapter.

In addition to confusion between ideas, invention, innovation and research, there is a distinction which is not always made between regarding innovation as a process (from an idea to a successful implementation) and innovation as an outcome, in the form of a product or service. The phrase "how to be innovative" in the title of this book refers to both developing the capability as well as the results.

Opportunities to learn

The wide range of organizations with a commitment to innovation offers opportunities to learn. Many companies will not be prepared to discuss the details of innovations under development, if this is commercially sensitive. However, they may well be prepared to discuss the processes by which innovation can be managed and stimulated, including the barriers encountered, and indeed such experiences are often presented at conferences concerned with the management of innovation. Direct contacts can also be made between non-competing organizations to discuss the innovation process.

Generating new solutions

It was noted above that major societal challenges require contributions from many stake-holders, including companies, government, and non-profit organizations. These bodies include a diversity of thinking, perceptions and goals. As discussed in Chapter 4, diversity of inputs is useful in the generation of new ideas; it is, however, valuable to first establish a shared perception of what innovation means, criteria for success, and factors which could slow or accelerate the journey from idea to success.

1.4 AIM, SCOPE AND USE OF THIS BOOK

The above has outlined the breadth and importance of innovation, and has indicated some areas of confusion. In light of this, the present book aims to provide a concise and useful guide to (i) concepts, (ii) practical applications and examples, and (iii) sources for further reading. As a result, readers will have a better idea of what it means to be innovative, and can thereby help their own organizations to become more effective in becoming innovative.

A short book is helpful for those with limited time to get to grips with innovation, and this is particularly true for those starting new jobs which involve innovation. New jobs can take up much time and energy, requiring knowledge to be acquired, skills to be developed, working relationships to be formed, and all in an initially unfamiliar environment. For this reason the present book aims to provide a concise guide, rather than a comprehensive coverage of all aspects of innovation, and this is further helped by three areas of focus.

The first area of focus is on early stage innovation, which is the process from an idea to Proof-of-Concept, (also referred to as proof-of-principle) and not the later stages of innovation. Early stage innovation has much in common across different sectors, whereas later-stage innovation, from Proof-of-Concept to successful implementation, is more business-specific.

The second area of focus is on scientific and technical innovation, bearing in mind the needs of scientists and engineers who take on a new role with respect to innovation, for example through moving from academic research to an employer with a commitment to innovation. Much writing on innovation is concerned primarily with business or strategic perspectives, and this is not directly relevant to those at the start of technical careers.

The third area of focus is on innovation by organizations, rather than by individual entrepreneurs. However much of the book may be of interest to the latter, bearing in mind that most start-ups aim to grow. While fast and informal decision making works well in a small group with a common goal, growth will increasingly demand more formalized processes. A common understanding of what innovation is, and how it can be achieved, provides a foundation for growth.

This book is based on the premises that efforts to be innovative are more likely to be successful if the following elements are in place:

— shared understanding of what innovation means.
— a process for defining areas for innovation, generating and prioritizing ideas, and allocating resources, as well as for making use of external resources (open innovation).
— developing behaviors, teams, communications and learning which support innovation.

These themes are dealt with as follows.

Chapter 1: Shared understanding

Given the breadth of innovation discussed above, and some of the confusion around the topic, there should be a shared understanding of what innovation means. Different parts of an organization can have different views on the subject, and on the criteria for determining success. Bringing an innovative idea to a successful conclusion (i.e. market impact) increasingly involves a range of external partners, as discussed in Chapter 3 (open innovation), and a common understanding of what innovation entails also plays a role in engaging with external partners.

The present chapter focuses on concepts involved in innovation. In addition to providing a basis for work within an organization, or with external partners, an understanding of the concepts also provides a framework for an interpretation of the constant flow of news items concerning innovation.

Chapter 2: A systematic approach

Recognition has grown that innovation benefits from being approached systematically, and not just left to chance. Some will argue that maximum freedom, plus sufficient funding, are important for researchers to come up with new ideas, but funds are rarely unlimited, and this implies making choices and setting priorities in areas to be investigated, and in considering the balance of risk and reward, as well as the balance of short term and long term rewards. The need for systematic approaches has grown as innovation increasingly becomes the task of teams which are multidisciplinary, multicultural, and globally dispersed. All the case studies presented in Chapter 5 relied on a structured process to develop the idea.

An article on "The discipline of innovation" published in 2002 described the benefits of a systematic approach (Drucker, 2002). Pinchot coined the term "Intrapreneurship" to describe the practice of working like an entrepreneur within an organization, and in 1999 wrote a handbook with Pellman on *Intrapreneuring in Action* (Pinchot and Pellman, 1999).

Chapter 2 describes a systematic approach from the first step of defining the scope for innovation up to the Proof-of-Concept. The benefits of a systematic approach generally outweigh leaving innovation to chance, but it must be emphasized that innovation is a messy business, influenced by feedback loops, personalities, and changes in ambition which may form part of the "spirit of the times" without necessarily being fully articulated. Any systematic approach adopted should be regularly reviewed and critically examined for opportunities to simplify it, and make it faster, cheaper and more effective. Reviewing existing processes against the ideas in this book, or from the literature referenced, or from new examples which emerge, will give ideas for improvements, and will help avoid the risk that an innovation process, once in place, develops into a rigid bureaucracy which becomes an end in itself.

Chapter 3: Open innovation

Chapter 3 discusses Open Innovation, which is about engaging with the external world, in order to achieve innovation more effectively, in terms of increased speed of implementation, or in reduction of costs, or through improved quality. This is essentially part of a systematic approach to innovation discussed in Chapter 2, but the topic has received much attention during the last 15 years and merits a separate chapter.

Chapter 4: Behaviors, teams, communications and learning

Chapter 4 covers the "soft" areas of behaviors, teams, communication and learning. These topics are not the natural habitat, at least at the start of their careers, of many scientists and engineers, but are nevertheless essential to the smooth and effective operation of groups of people, ranging from small teams to multinationals. These topics are particularly important in the early stages of innovation, since it is here that uncertainties of the outcome, both as costs and benefits, are at a maximum. This uncertainty, when combined with the drive and enthusiasm of an innovative mindset, can lead to conflicts, and a balance to be addressed here is that between enthusiasm and objectivity. The importance of these factors is illustrated by the prefatory quotes — about five hundred years apart — from Machiavelli and Negroponte.

Chapter 5: Case studies

The ideas covered in Chapters 1–4 are illustrated with eight case studies, six of which are technical in content. These case studies are based on individual accounts, which thereby give an authentic view of early stage innovation in practice. The case studies also provide an opportunity to learn from other practices, and a comparison across the case studies is given.

Chapter 6: Next steps

Chapter 6 discusses some myths and pitfalls, provides topics for discussion, and indicates some likely developments in the management of innovation. There is little doubt that new examples, insights, and approaches will continue to materialize.

Use of this book

The content of this book aims to give the reader a better appreciation of how he or she can be more effective in innovation (producing innovations and developing the required skills), based on the ideas and practical examples. Much of the book can also be seen as a basis for discussion, on what could be implemented, and on what is already effective or could be improved.

References to a significant number of books and articles are given, and this allows deeper exploration of the many facets of innovation which are touched on, as well as providing an entry point to further literature. More comprehensive treatments of innovation can be found, for example, *Making Innovation Work* (Davila, Epstein and Shelton, 2012); *Innovation and Entrepreneurship* (Bessant and Tidd, 2007); *Business Innovation for Dummies* (Hiam, 2011); *Innovation Management and New Product Development* (Trott, 2011); and *The Oxford Handbook of Innovation* (Dodgson, Gann and Philips, 2015). These books are several hundred pages long, and the last two are over six hundred pages.

Not all of the present book may resonate immediately with the reader: different sections of the books may assume more meaning as experience and careers develop. Growing experience of projects, and of different roles, (for example technical, commercial or managerial) will provide material which can be used to illustrate, extend, or challenge the ideas presented here. It is hoped, however, that this material forms a foundation on which to build. The remainder of this chapter deals with some concepts.

1.5 DEFINITIONS

The word innovation is in common use, and many people feel they understand what is meant by it, without necessarily being able to define it. Many definitions can be found. One author observed that he had written several books on innovation without defining the term. He referred to a list of 25 often contradictory definitions before settling on "innovation is something different which has impact," in which "different" is in the eye of the beholder and impact "means some kind of measurable result" (Anthony, 2012a). Other definitions include "The effort to create purposeful, focused change in an enterprise's economic or social potential" (Drucker, 2002) or the OECD (Organization for Economic Cooperation and Development) definition "An innovation is the implementation of a new or significantly improved product (good or service), or process, a new marketing method, or a new organizational method in business practices, workplace organization or external relations" (OECD, 2005).

These examples emphasize different aspects of innovation: that it involves effort, change, newness and outcome. It perhaps does not matter much which definition is used, provided it is applied consistently, and that limitations of the definition are recognized. For example, the definition "innovation is invention plus application" may be good for technical organizations in that it emphasizes that an invention alone is not an innovation, but it restricts thinking in wider terms such as business model innovation or social innovation.

The above definitions are not succinct or memorable enough to be useful in everyday conversations aimed at achieving clarity on what an innovation is. From this point of view a good definition which will be used in the rest of this book is:

"Innovation is the successful exploitation of new ideas"

This definition (Dodd, 2008) has the advantage of being short, easy to memorize, and encompassing a result and a process.

Note that this definition can be equally well applied to both commercial and non-commercial innovations. The definition means that it is superfluous to refer to a "successful innovation," since innovation is defined as being a success. So a new idea, an invention, a prototype, or a technology demonstration does not count as innovation. They are, however, steps on a journey towards successful exploitation, which would then justify the use of the term innovation.

A new business is not necessarily innovative. Opening another branch of a fast-food restaurant chain, with nothing particularly new in location, customers or menus, would not count as an innovation.

Distinguishing between ideas, inventions, prototypes, technology demonstrations and innovations is necessary to avoid confusion about meeting the ambition of being innovative. A shared understanding of what innovation is can only help an organization become more effective in being innovative.

Examples of innovations can easily be found. As mentioned above, the average household contains many appliances (refrigerator, freezer, air-conditioning, heating, lighting, television, radio, etc.) which meet the above definition, as do social innovations such as microcredits, and developments in new energy technologies such as solar panels or offshore wind turbines.

As for failures, a "Museum of Failure" opened in Sweden in 2017. The website notes that the majority of all innovation projects fail; and that the "collection consists of seventy failed products and services from around the world. Every item provides unique insight into the risky business of innovation" (Museum of Failure, 2017). Displayed items include the Apple Newton, Google Glass, Nokia N-Gage, a Kodak digital camera, and Lego Fiber Optics.

1.6 HISTORICAL NOTE

Innovation as a subject tends to produce enthusiasm for present and future developments, and not all writing on innovation contains a sense of historical perspective. While in no way comprehensive, the following gives a few marker points.

The history of the human race is a story of a stream of innovations. A list of innovations from 1400–2000 started with double entry accounting (1300–1400), the printing press (1440) and the concave lens (1451), but this list (Johnson, 2010) could be extended much further back in time: it has been pointed out that "the control of fire some 300,000 to 500,000 years ago is one of the earliest, and possibly one of the most important innovations of mankind" (Verloop, 2004).

The study of innovation is much more recent and the recognition that innovation is a key economic driver is credited to Schumpeter (1883–1950), an Austrian economist and later professor at Harvard. The last 35 years have seen interest developing in a systematic approach to innovation. In 1982 *Readings in the Management of Innovation*, a collection of about 45 articles, was published with the aim of providing a "unified and comprehensive view of the problem and some possible solutions to the management of innovation" (Tushman and Moore, 1982). In 1984 the Center for Innovation

Management Studies was created at Lehigh University (it is now located in North Carolina State University's Poole College of Management), with the goal to better understand the technological innovation process and how it could be managed more effectively. The 1985 best-selling book *Innovation and Entrepreneurship* by Drucker was seen as the first book to present innovation and entrepreneurship as a purposeful and systematic discipline (Drucker, 1985).

Writing in 1999, in an article which was an update of an earlier article in 1992, Galbraith asked, "If everyone is in favor of it, why is innovation so difficult to master?" (Galbraith, 1999), stating that the question had been addressed continuously for at least 30 years but that peaks of interest had occurred in the late 1960's, and again in the 1990's.

The "dot com" era of the late 1990's saw an upsurge of interest in radical innovation, as illustrated by the book *Leading the Revolution* (Hamel, 2000), and an article by the same author, "Bringing Silicon Valley Inside" (Hamel, 1999), which focused on how major companies could incorporate the spirit and dynamism of start-ups into their large organizations, as well as by the book, *Radical Innovation: How Mature Companies Can Outsmart Upstarts* (Leifer *et al*, 2000).

One view of the history of innovation is that it can be divided into four eras (Anthony, 2012b): firstly, the era of the lone inventor, up to about 1915, and illustrated by Gutenberg (printing press), Edison (light bulb) or the Wright brothers (aeroplane); the second era is that of corporate laboratories such as those of DuPont (nylon), Proctor and Gamble (Crest, Pampers, and Tide brands) or Lockheed Martin (the U2 spy plane, the SR-71 Blackbird fighter jet); the third era is that of start-ups backed by venture capital companies, with Apple, Microsoft, Cisco Systems, Amazon, Facebook and Google all receiving support from VC companies such as Kleiner Perkins Caufield and Byers, or Sequoia Capital; and the fourth era is seen as that of the "Corporate Catalysts": individuals working for large companies who are able to tackle big problems and who assemble internal and external resources to do so. Examples resulting from this approach include Healthy Heart for All (cardiac diagnostic services and financing developed by Medtronic in order to bring pacemakers to people who previously could not afford them), Pureit (a portable water purification system developed by Unilever, with the aim of supplying clean water to

500 million people), Uwezo (crop-protection chemicals and seeds, distributed in sachets, and supported by education and training, developed by Syngenta) and Smarter Cities (a combination of technology and services to help cities manage energy, water, traffic, parking, public transport and crime), developed by IBM.

Books on systematic approaches to innovation include those by Verloop (2004), Anthony (2012a) and Markham and Mugge (2015). More general accounts of innovation are provided by Johnson in his books, *Where Good Ideas Come From: The Seven Patterns of Innovation* (Johnson, 2010), *The Innovator's Cookbook*, which is a collection of articles and interviews (Johnson, 2011), and *How We Got To Now: Six Innovations That Made The Modern World* (Johnson, 2014).

1.7 REFINING THE CONCEPT OF INNOVATION

Exploring the definition

The definition given above, "innovation is the successful exploitation of new ideas," can be developed by thinking about what "new" means and what "success" means (the question of "what is an idea" quickly gets into a philosophical discussion which is beyond the scope of this book).

"New" has two implications with respect to innovation. Firstly, "new to whom?" and secondly, "how new?"

New to whom?

For scientific discoveries, "new" in academic terms is related to the question of whether a result or insight is sufficiently interesting to merit publication, through increasing degrees of recognition, ranging from invitations to speak at conferences up to award of a Nobel Prize. However, for a patent lawyer scientific newness may not be the basis of obtaining a patent, for which criteria are terms such as novel, useful and non-obvious (exact terms depend on different patent-granting authorities). Here novel means "not previously described" and this is not necessarily significant scientifically. Indeed, "non-obvious" may be in conflict with the logic behind a scientific discovery. So "new" for the Nobel Prize committee has different implications than for a patent-granting authority. A business unit may not be concerned if the finding is scientifically new, or can be patented, if there are other ways to make a commercial success of the finding, for example through combination with other, proprietary, technologies, or through unique access to markets, or through exclusive partnerships, or through being first to market (the "first mover" advantage). Finally, users of products containing innovative technologies may only be interested in what the product or service can do, and not in the new scientific or technical advances which lie behind the features.

The point here is that when it is claimed that a finding is "new" it is always worth exploring the question of "new to whom?" This is relevant for those who have recently completed academic research, with a focus on scientific newness, and have to deal with the different perspectives of patent and business departments. These distinctions are also relevant in establishing collaborations between universities and industry, and this topic is explored further in Chapter 3, which deals with Open Innovation.

How new?

A second aspect of "new" is the question of "how new?" which can refer to time (how long ago was it invented?) or to the degree of newness. Here the focus is on the latter aspect, and this requires discussion of the terms incremental, radical and disruptive innovation, which have become widely used in recent years. It may be asked why these distinctions matter. The answer is that these types of innovation can involve different approaches to managing risk, different strategies, or different organizational forms. It should be realized that the distinction between incremental and radical innovations is not a binary choice and examples in practice fall on to a continuous scale. The case studies in Chapter 5 include several examples of innovations which rely on significant extensions to prior practice, and are not simply incremental or radical.

Incremental innovation

As the name suggests, incremental innovation refers to relatively minor changes to existing products or services. Since the markets are known, the risks — and rewards — are smaller than for radical innovations. This is both a strength and a weakness: there is less chance of making a disastrous investment, but a focus on incremental innovations may encourage a culture of risk avoidance, which may result in an unpleasant shock caused by competitors willing to take bigger risks. This leads some writers and consultants to advocate a focus on radical innovation. However, the cumulative effect of a steady series of small improvements can be very significant, with examples including developments in computing power (Moore's law), and on-going progress in the performance, and in reducing the cost, of wind turbines and solar cells.

Radical innovation

Radical innovation refers to a major step-out. For companies this includes significantly new technologies or markets. Examples include the introduction of digital photography, provision of e-learning such as MOOCs (Massive Open Online Courses), and successive waves of new products in providing recorded music: the 78 rpm gramophone record, the 33 rpm LP, the 45 rpm single, the cassette, CD and DVD, as well as downloading and streaming services. Each of these products involved new technologies and in some case new markets: the cheap transistor radio, coupled with developments in popular music, opened a new market of teenagers in the late 1950's.

Developing new markets and new technologies at the same time involves dealing with two sets of risks, and tackling both at the same time will put a strain on organizational capacity. Unless there is a compelling reason to do both at the same time, and it is possible to do this in terms of procedures, decision points and internal communication, it is better to decide if development of the market or development of the technology should have the highest priority.

With respect to organization of radical innovation, the model adopted by the Pentagon's Defense Advanced Research Projects Agency (DARPA) is of interest. This agency has "arguably the longest-standing, most consistent track record of radical innovation in history" (Dugan and Gabriel, 2013). The innovations, for which the US military was the original customer but which were the basis of many new and major industries, include the Internet, RISC computing, global positioning satellites, stealth technology, unmanned aerial vehicles (drones) and MEMS (micro-electro-mechanical systems) used in airbags, ink-jet printers and video games.

The DARPA model is based on three elements: ambitious goals, which include a commitment to working in Pasteur's quadrant (described later in this chapter) which the authors describe as "pushing the frontiers of basic science to solve a well-defined, use-inspired need"; temporary project teams, which are highly focused and of relatively short duration; and independence with respect to selecting and running projects. At the time of the article, the authors were engaged in implementing the DARPA model in the private sector, at Motorola Mobility, which was later acquired by Google.

In summary, there are two points here: firstly, when faced with the possibility of developing a radical innovation an organization should consider whether it can manage the development of the technology or service involved at the same time as developing a new market, or whether these two aspects should be prioritized with respect to each other. The second point is to consider whether the organization should be adapted, for example to follow the elements in the DARPA model described in the preceding paragraph.

Disruptive innovation

The term disruptive, like the word innovation, is so widely used that it leads to confusion. Disruptive innovation was a term introduced in 1997; it is now frequently used for any significant transformation. As originally presented (Christensen, 1997), the term referred to the process in which established businesses are threatened by newcomers, in one of two ways. Either newcomers establish a new market, or they establish a foothold at the low end of an existing market. In both cases, by steadily improving the performance of their products or services, they are able to challenge established competitors. More recently developments in the thinking about disruptive innovation have been reviewed (Christensen, Raynor and McDonald, 2015). One example of disruptive innovation concerns the company Xerox, who pioneered photocopying, and who focused on large corporate customers. The market for personal photocopiers, encompassing small businesses, home offices, clubs, libraries and other customers was neglected. New entrants who started in these markets developed into significant competitors to Xerox in the mainstream photocopying markets.

In contrast, Christensen argues that the taxi service Uber is not disruptive, in the sense that Uber did not start in the low end of a market, nor did they target people who did not use taxis (non-consumers).

What does "successful" mean?

The "successful" exploitation of a new idea means that customers buy, or adopt, the product or service. The numbers required to constitute a success clearly depends on the particular market: a very high technology piece of equipment, for example a new medical scanner, may have very limited sales but still meet the definition of success, whereas the manufacturer of an innovative automobile model will require much higher numbers. Market share is also a valuable metric. The time scale needs to be established: some innovations may be very successful in a short period, but become less significant, or obsolete, over a longer period.

Stakeholders in developing an innovation will have an idea of what constitutes a success, but it is helpful to share these perceptions at an early stage and clarify any differences.

1.8 HOW INNOVATIONS START: OPPORTUNITIES, NEEDS, PROBLEMS, ADAPTATIONS AND COMBINATIONS

There are many ways that innovations start, and this section discusses some of them. An additional source of innovation, research and development, is covered in the following section.

Opportunities, needs and problems

Those motivated by innovation easily become enthusiastic about new ideas for products or services. This enthusiasm for such an opportunity is reinforced when it becomes apparent that the idea can also be carried out. The saying "Build a better mousetrap, and the world will beat a path to your door" (attributed to the American poet Ralph Waldo Emerson) is, however, a poor guide to innovation since one element is missing: the need. People may not wish to pay more for a better product unless it addresses a need.

The belief that superior features or performance will alone be sufficient to attract customers is sometimes known as the "better mousetrap fallacy."

Indeed, as has been noted, "the starting point of an innovation is an idea of a need that isn't being met, coupled with an idea of how it could be met" (Mulgan, 2011). So sketching out new opportunities should be balanced with an analysis of potential customers: who would be interested in purchasing or adopting the offering, what need is it meeting, or what problem does it solve, and why would this particular offering be better than others?

Some needs or problems are very clear, for example, the need to provide energy which is affordable and which has acceptable environmental impact. Some needs are expressed by customers who desire particular features in a product or service, and useful inputs for developing innovations can be obtained from interviewing customers. However, this "voice of the customer" is but one input, and customers may not always know what they want. Henry Ford is quoted as having said "If I had asked people what they wanted, they would have said faster horses," but there is in fact no evidence that he said this (Vlaskovits, 2011). Nevertheless, the quote does capture the fact that customers may not understand their own needs. Steve Jobs, the founder of Apple, made the same point by saying: "But in the end, for something this complicated, it's really hard to design products by focus groups. A lot of times, people don't know what they want until you show it to them" (Jobs, 1998).

Christensen has pointed out that Sony's introduction in 1955 of a cheap transistor radio was successful because it met a need, amongst American teenagers of the time, to have an affordable device which would enable them and their friends to listen to new music such as rock and roll out of earshot of their parents. Quality was of lesser concern. It is unlikely that these teenagers, prior to 1955, were sitting around saying "I wish we had a cheap, portable transistor radio." In other words, the need was there, but not articulated (Christensen and Raynor, 2003). Such unarticulated needs are sometimes referred to as latent needs.

An earlier example of unarticulated needs which almost destroyed a business venture has been described in the case of "Murex and the red tins" (Verloop, 2004). The "Murex" was the name of a ship launched in 1892 and the first of a fleet built for M. Samuel and Co., which later became part of the Shell group of companies. The "Murex" was designed to supply kerosene (derived from crude oil and then used for heating and lighting) in bulk, and more cheaply, to Far Eastern markets, replacing the delivery of kerosene in tins. The design of a bulk oil carrier had to meet stringent safety requirements; in addition the tanker could be steam-cleaned, which meant that other goods could be brought back to Europe from the Far East. The intention was that customers would keep their old blue tins and regularly fill them with kerosene delivered by tanker. However, this overlooked the fact that the empty tins had many additional uses for customers, including roofing, birdcages, hibachis and tea strainers, and the value this represented was more important than the lower cost of the kerosene. The new business venture built on modern ship technology and bulk delivery was saved by manufacturing bright red tins locally which the customers could use. The new colour stood out, and helped to create a competitive advantage.

Analysis of problems may or may not lead to innovation. For example, when faced with a device or machinery which does not work, a solution may simply consist of identifying and replacing a defective part. Understanding why and how the part failed, however, could well lead to an innovation in the form of an improved design. This may require tracing sequences of cause and effect, an approach known as root cause analysis. It can also be valuable to "reframe" the problem: exploring alternative perspectives on, and definition of, the problem before proceeding to generating solutions. An example is dealing with irritation caused by waiting for a slow elevator (Wedell-Wedellsborg, 2017). A conventional problem analysis would suggest upgrading or replacing the elevator to make it faster. Instead of defining the problem as "the elevator is too slow" the problem can be reframed as "the wait is irritating" or "the issue is peak demand." These different problem definitions lead to different sets of solutions. The article describes seven approaches to reframing problems and finding creative solutions: establishing legitimacy (getting support for the reframing approach); bringing outsiders into the discussion (see also the discussion in Chapter 4 on diversity of opinions); getting people's definitions in writing (to get clearer formulations than opinions expressed in meetings); asking what is missing in the description of a problem; consider multiple categories (what is the type of problem?); analyzing positive exceptions (identifying situations where the problem did not occur); and questioning the objective. The article notes that reframing is not easy and quotes a senior executive from the defence industry: "I was shocked by how difficult it is to reframe problems, but also how effective it is."

Balancing opportunities with needs, listening to customers, and exploring unarticulated needs are parts of a systematic approach to innovation and these aspects are discussed further in the following chapter.

Adaptations

A common source of innovation is adapting solutions or knowledge in one field to problems in other areas, and this is can be seen in some of the case studies discussed in Chapter 5. One area of current interest is the subject of biomimicry, which aims to look to nature for solutions or inspiration. One of the best known and commercially successful examples is that of Velcro, the fastener. The idea for this came from a Swiss engineer who, after a hunting trip in the Alps in 1951, wondered why his dog was covered in burdock burrs. Examining a burr under a microscope revealed many tiny hooks, and this was developed, over a number of years, into Velcro, the first patent being filed in 1952.

Combinations of elements

It is tempting to look for a single cause or discovery behind an innovation. However, reality tends to be messy, and there are often many complementary contributions necessary to bring an idea successfully to market. A technical discovery will require development, a business model, and a way of protecting the innovation from competitors. Disentangling the many individual contributions in an organization where roles and strategy evolve, or completely change, is difficult. In addition, the account given by participants of their role in developing an innovation tends be colored by whether the project was successful or not.

Even at a technical level, there can be many contributing and separate technologies. There are 12 major technologies integrated within the Apple iPod, iPhone and iPad that standout as enablers or differentiators (Mazzucato, 2014):

1. Microprocessors or central processing units (CPUs)
2. Dynamic random-access memory (DRAM)

3. Micro hard drive storage or hard drive disks (HDD)
4. Liquid-crystal displays (LCDs)
5. Lithium-polymer and lithium-ion batteries
6. Digital signal processing
7. Internet
8. Hypertext transfer protocol (HTTP) and hypertext markup language (HTML)
9. Cellular technology and networks
10. Global positioning system (GPS)
11. Click-wheel navigation and multi-touch screens
12. Artificial intelligence with a voice-user interface program

Mazzucato discusses the origin of these technologies with respect to the role of the state. This list also shows that innovation can result from the ability to recognize the potential of combining separate technologies. The rapid development of technologies in many fields, as well as the increasing number of ways to share knowledge and collaborate, suggest that new, and perhaps surprising, combinations of technologies will be the source of much innovation in the future. Creating new ideas through combinations of skills, competencies or technologies is discussed in Chapter 2.

1.9 INNOVATION AND R & D

The links between research and innovation have been much discussed by economists and policy makers, and many practicing scientists may be surprised by the amount which is written by others about their work. The following summarizes some of the ideas.

A classic view of technological innovation is that basic research leads to new ideas which undergo development into a technology, followed by commercialization. This view is frequently thought (but incorrectly, see below) to be derived from the 1945 report, "Science — the endless frontier" written by Vannevar Bush (no relation to later presidents) and commissioned by President Franklin D. Roosevelt in 1944. Roosevelt wished to establish how science after World War II could build on the scientific efforts which had led to the development of the atomic bomb, radar and penicillin.

The Bush report distinguishes between basic research (a term coined by Bush and defined as a contribution to "general knowledge and the understanding of nature and its laws") and applied research, which focuses on practical application of research. In addition to the distinction between basic and applied research, an important idea was that basic research was the principal source of technological innovation, and that there was a linear progression from basic research to applied research to development to commercialization. This report argued for strong government support of "basic research" in the post war period. In a retrospective essay published in 2010 (65 years after the original report) it was observed that the phrase "basic research" could be used to imply both unrestrained fundamental or curiosity-driven science, as well as science which would form the basis of innovation (Pielke, 2010). This term clearly struck a chord: Pielke reports that the term "basic research" was used 4 times in The New York Times in 1944, rising to a maximum of 159 times in 1957. Since then the terms basic and applied research have become widely familiar.

A further step in thinking about science and innovation was made in the book *Pasteur's Quadrant: Basic Science and Technological Innovation* (Stokes, 1997). This book reviewed the origin and impact of the ideas in the Bush report, noting that "the belief that scientific advances are converted to practical use has been a staple of R & D managers everywhere." Stokes discussed three flaws in this view: (i) that the linear model is too simple, (ii) that not all technological innovation is rooted in science, and (iii) that the flow can be in both directions: technological developments can drive basic science. Stokes went on to give an alternative classification of scientific output in terms of two axes, the quest for fundamental understanding, and the consideration of use. The resulting four quadrants (Fig. 1.1) were exemplified by Bohr (a quantum physicist, who won the 1922 Noble Prize) as an example of contributions high in scientific excellence and low in practical relevance; Edison, the famous inventor whose research and development led to major new industries based on electric light, motion pictures, and sound recording, and who is an example of focus on practical relevance; Pasteur who made many contributions to both the causes and prevention of diseases, and who is seen as a leading founder of bacteriology, and thus

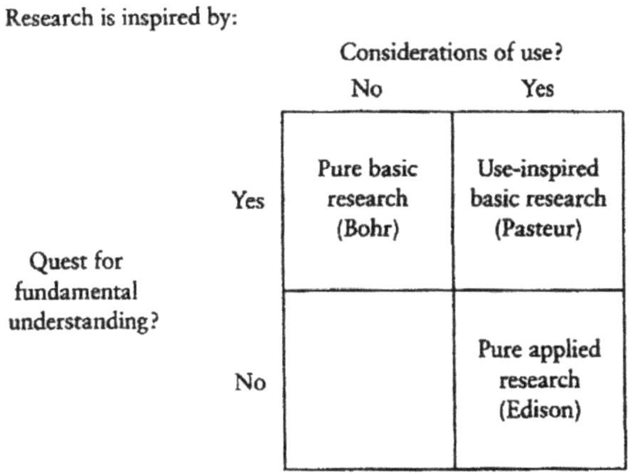

Fig. 1.1. The quadrant model of scientifc innovation. Reprinted with permission of the publisher. (From *Pasteur's Quadrant*, by D.E. Stokes, copyright ©1997 by The Brookings Institution, Washington, D.C.)

exemplifies both scientific excellence and practical application. It should be clear that this classification refers to the primary motivation at the time of the research, and fundamental work focused on understanding can lead to later practical advances. A second point is that the quadrants are not mutually exclusive: individual scientists, research organizations, and those providing funding can make choices regarding the amount of effort made in each of the quadrants.

As regards the fourth quadrant, it has been said that "Stokes was tactful enough not to label the lower left corner, where the research neither advances research frontiers nor addresses significant practical problems. The fact that a majority of scientific papers are never cited by anyone other than the author (or his students) suggests that this box is not entirely empty" (Elg, 2014). While elegantly formulated, this view does not do justice to Stokes, who stated that this fourth quadrant includes research that "systematically explores particular phenomena without having in view either general explanatory objectives or any applied use to which the result will be put"; an example given by Stokes is the systematic research into the markings and incidence of bird species. Stokes also remarked that research in the fourth quadrant can be used to develop skills which are valuable precursors to advances in the other quadrants.

With respect to the linking of science and technology, Stokes wrote that the one-dimensional, one-way model needed to be replaced by a picture in which both science and technology develop semi-autonomously, with strong or loose interactions, and in either direction. In addition to developing these ideas, the book discusses policy and societal implications.

Despite the widespread attribution of the linear model of innovation to Vannevar Bush, it has been shown that the linear model of innovation predates the Bush report, that it has long been criticized (since the 1960's), and that alternatives which have been proposed to the linear model have had little impact (Godin, 2006). This author suggests that the widespread use of the linear model, from basic research through to commercialization, is due to the production of official statistics on the steps in the model, and the fact that quantitative data for alternative models are not available to the same extent. The same author has since documented and reviewed the history of models of innovation, of which the linear model is the most important (Godin, 2017). The models discussed in this book fall into two broad categories: process models, in which time plays a role through a sequential series of steps in innovation, and system models, which focus on the organizations which play a role in innovation (for example companies, universities and governments) and how they interact.

Criticism of the linear model of innovation was also made in a publication reviewing forty years experience with understanding the links between innovation and research. It was stated that the model of "Basic research > Applied research > Development > Production > Market" gives a poor description of the relationship between research and innovation, and that studies before the Second World War emphasized the "relationship between scientific discovery and application is a complex interaction which cannot be simplified into a one-way information flow" (Elg, 2014).

An illustration of how the linear model of progression from basic research to application does not apply is provided by an example of applied research leading to fundamental insights. In solving a very practical problem of interference by weak radio waves, in detecting the signals obtained by bouncing radio waves off balloon satellites, Penzias and Wilson discovered in 1964 cosmic microwave background radiation. This provided strong support for the "big bang" theory of the origin of the universe (in contrast to the then competing model of a steady state universe); Penzias and Wilson were awarded the Nobel Prize for physics in 1978. Penzias and Wilson worked at Bell Laboratories in the USA, one of the most important laboratories to combine both scientific excellence and practical relevance. Major contributions were made to the award of eight Nobel prizes, as well as to the development of the laser, the transistor, radio astronomy, information theory, the operating system UNIX, and the programming languages C and C++, all of which have found major and widespread application. An account of Bell Labs and the interplay of science, technology, business, the management of innovation, and personalities has been written (Gertner, 2012).

Those working in research organizations which cover the whole spectrum from basic research to support for commercialization and technical service to customers after commercialisation will recognize that innovations can start anywhere in this chain, from new scientific insights or discoveries in basic research, from wide-ranging exploratory research, from spin-offs from development work, from commercial awareness of new opportunities requiring a technical advance, or from problems encountered by customers.

Since Stokes' book in 1997, and the heyday of Bell Labs, there has been much interest and activity in developing the capability to commercialize the results of research. The increase in science parks, technology transfer offices, entrepreneurs, incubators and venture capitalists focused on technology illustrate this activity, and will likely lead to new models of the relationship between R and D and innovation.

Apart from the question of the linear model, the belief that R and D and innovation are clearly linked is exemplified by the opening sentence in a recent report on innovation in relation to US manufacturing: "If any one factor explains why the US has remained the world's innovation powerhouse since World War II, it is the country's overwhelming leadership in research and development" (Sirkin, Rose and Choraria, 2017). This view is reinforced in popular culture by a statue in the "American Adventure" pavilion at Disney's Epcot theme park in Florida. This statue (Fig. 1.2) bears the title "The spirit of innovation" and the figure portrays a scientist carrying out an experiment, complete with white coat, flask and test-tube. Many visitors will have left with an impression that R and D and innovation are intimately connected.

Fig. 1.2. Spirit of Innovation — or the Spirit of Science? (© Disney. Reproduced by permission.)

However, as described above, the links between R and D and innovation are complex, and that simply funding more of the former in the expectation that more of the latter will result is not justified, or at least not in general. Mazzucato, in writing about innovation-led growth, discusses various myths, or false assumptions, about the drivers of innovation, including "Myth 1: Innovation is about R and D." The author notes that studies of different companies have shown that there are examples of a positive impact of R and D on growth, but also examples of no significant impact, or a negative impact, when R and D is only a cost. There are also differences between sectors, and between particular phases of an industry life-cycle

(Mazzucato, 2014). It is clear that different sectors have different degrees of need for research to generate or support innovations: services for example require less research than the introduction of new energy technologies.

Serendipity

Although science is generally considered to be methodical, controlled and logical, many scientists will recognize the role played by "accidental" and "lucky" observations, although these may not be reported as such in publications. The word serendipity is often used for accidental discoveries, but the original description of serendipity (in a letter written in 1754 by the English writer Horace Walpole) was of "Three princes of Serendip making discoveries, by accidents and sagacity, of things they were not in quest of." It has been noted that sagacity — defined as "penetrating intelligence, keen perception and sound judgement" — is often left out of modern definitions of serendipity, but that it is a key element (Rosenman, 2002). This is also reflected in Pasteur's statement "Chance favors the prepared mind." Rosenman states that "serendipity has played a crucial role in science," with examples including, amongst many others, the discovery of penicillin and the microwave oven. The role played by unplanned elements is nicely captured by Koestler's view that "The history of discovery is full of arrivals at unexpected destinations and arrivals at the right destination by the wrong boat" (Rosenman, 2002).

Summary

The above discussion makes clear that the links between R and D and innovation are complex, and not adequately covered by a simple — and widely used — linear model of technological innovation, which starts in basic research and proceeds to commercialization. Awareness of the ideas and literature cited above will help those working in the early stages of technical innovation to reflect on how innovation in their own particular area actually takes place, and this understanding will help develop better practices.

2 Building the Innovation Process

2.1 INTRODUCTION

Many people have ideas for new or better products or services, as well as for improvements to their own organizations. However, without some way to deal with these suggestions, they are likely to remain as jottings in a drawer, or languish in a rarely-opened suggestion box. There is thus a need for a systematic approach to handle new ideas effectively; in other words, to build the innovation capability. This capability has two important components: (i) a process for generating, selecting and developing ideas, and (ii) a culture which is supportive of innovation. In this chapter we look at the first of these factors, and in Chapter 4 at the second one.

The approach discussed in this chapter can be used at different scales, for example within a project team, within a department, or across an organization. An integrated system is presented, but it may be appropriate to make use of specific parts, in order to improve the aspects which are weakest in a particular working environment.

Lists of innovations typically include many technological advances, and sometimes social innovations such as long-distance learning or micro-credits. Few people, however, would quote Peter Drucker, who wrote: "Management of business processes is one of the great innovations of the 20th century" (Drucker, 2001). Innovation itself is sometimes thought of as not requiring any management, and that freedom, time and money will be sufficient. This might be true for individuals working largely on their own, at least in the early stages of innovation, but taking the idea all the way through to commercialization will require other people and some form of managed process. In addition, multidisciplinary, multicultural teams spread around the world benefit from the outset from processes which effectively combine individual contributions and which maximize chances of success. The realization that innovation is a business process which can, and should, be managed has grown in recent years, and all the case studies presented in Chapter 5 make use of structured processes to develop an innovation.

2.2 BUSINESS PROCESSES

All large organizations rely on procedures, or business processes, in many areas such as safety, recruitment, promotions, strategy, resource allocation and business monitoring. Collectively these are regarded as "the bureaucracy," and often seen as a necessary evil of corporate life. It is hard to imagine how things would work without such processes, but complaints about the time and complexity involved are common. It is somewhat paradoxical that scientists and engineers who are highly motivated to improve technical processes frequently regard business processes with less enthusiasm, while in both cases the quality of the process contributes to the quality of the outcome. Posing the question "Why is this manufacturing process not working properly?" is likely to meet with more interest than "Why is this meeting not working properly?"

Small organizations can pride themselves on their lack of bureaucracy, and ability to make decisions quickly. While this leads to an effective and satisfying work environment on a small scale, it should be remembered that with growth the need for business processes quickly becomes apparent. What might be effortless with five people becomes more difficult with fifty or five hundred. Small companies are usually focused on growth, but if successful face the challenge of developing fit-for-purpose processes while at the same time expanding the business.

With respect to designing or improving a process, a balance needs to be maintained between standardization across all parts of an organization, and the flexibility for individual units to develop systems as they see fit. The former approach can lead to centralization, rigidity and an approach which is not fit-for-purpose; on the other hand allowing each unit or team complete freedom can lead to a multitude of inconsistent definitions, criteria, and views on what constitutes success. The balance can be achieved through deciding which topics should be common, for example a definition of innovation, or having the same basic process in place, but allowing for differences where necessary. For example, one unit may, explicitly or implicitly, only consider radical innovations to be significant, whereas another part may be focused on a stream of incremental improvements. Formulating, and making known, the criteria for success, will allow better recognition of the fact that different parts of an organization can have different perceptions of what constitutes an innovation.

The difference between standardization of processes, and allowing more room for judgment, has been explored in an article entitled "When should process be art, not science" (Hall and Johnson, 2009); here "art" is taken to mean "judgment-based work," "craft work" or "professional work." The processes discussed which merit a more artistic approach include leadership training, business development and industrial design, all of which can impact innovation: with respect to business development the authors note "spotting new opportunities and envisioning how the business could exploit them can't be reduced to a formula."

Business processes are often developed into complicated flow diagrams. Reality, however, is considerable more messy, as illustrated in Fig. 2.1. The ability to manage the innovation process requires the ability to strike a balance between the messiness and the formal process. Too much messiness leads to inefficiency, while too much process becomes cumbersome. It is worth reflecting from time to time whether a given way of doing things — a process — would benefit from more, or less, structure and detail.

In considering processes, it is also worth reflecting on how the processes in place came about. Frequently they are regarded as part of the furniture, and not given a second thought. Processes can be traditional ("this is the way we have always done it"), evergreen (updated in the light of experience), designed for a specific goal, or emergent (start on a small-scale and evolve). The difference between the last two is illustrated by paths in a park or on a campus, which have been carefully designed and laid-out, only to find that users create their own shortcuts. So keeping some space and freedom to experiment before finalizing procedures (or pathways) is valuable, but must be balanced against the need to deliver results.

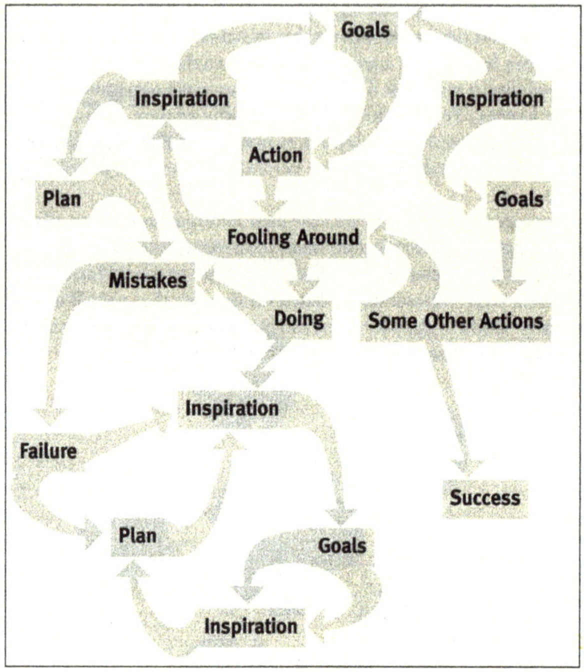

How Innovation Actually Happens

Fig. 2.1. How innovation actually happens. Reprinted with permission of the publisher. (From "Intrapreneuring in Action", copyright © 1999 by G. Pinchot and R. Pellman, Berrett-Koehler Publishers, Inc., San Francisco, CA. All rights reserved. www.bkconnection.com.)

2.3 AN INNOVATION PROCESS

While innovation generally sounds like fun, and a good thing, the journey from a fragment of an idea, or even a blank sheet of paper, to a successful business is long and hard, with many opportunities for frustration, disappointment and failure. Key steps along the way are:

- generating ideas
- turning them into potential opportunities, at which point Proof-of-Concept is established
- developing a detailed business case

- full demonstration of the idea
- launching a successful new offering, improvement, or saving

These steps are shown in Fig. 2.2. In moving through this sequence, the uncertainty is decreased, and the costs increased. At the stage of the initial idea, the costs are low but there is considerable uncertainty about whether the idea is viable. At the end the uncertainties have been largely dealt with, but (major) costs have been incurred.

Various books have described the whole process, and in some cases include detailed checklists and worksheets (Pinchot and Pellman 1999; Verloop 2004; Markham and Mugge 2015). Consulting companies also provide idea management systems (e.g. Imaginatik). The approaches can vary in detail, but in general encompass the above steps. For example in a book on *Managing Innovation by Understanding the Laws of Innovation* Verloop refers to the three fundamental stages of (i) generating the idea, (ii) demonstrating the concept, and (iii) investing to extract benefit (Verloop, 2004). Markham and Mugge (2015) present a detailed, practical

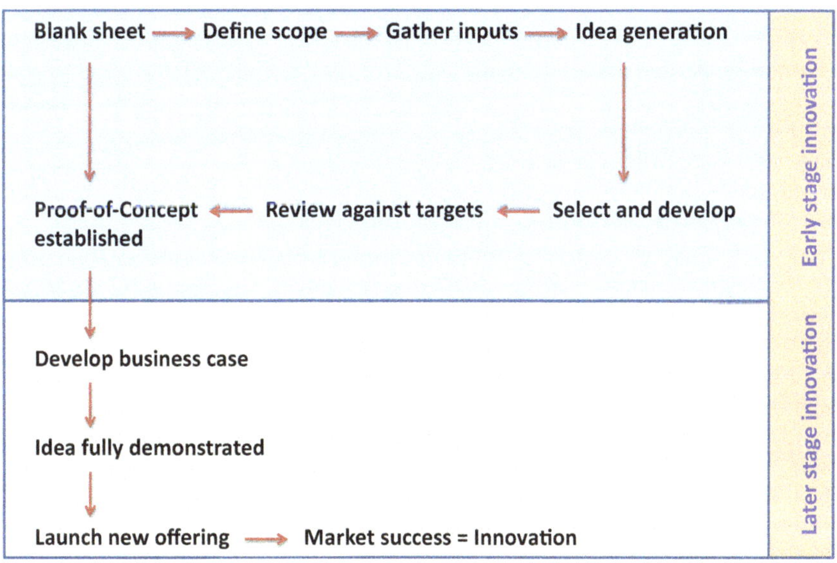

Fig. 2.2. Early and later stage innovation.

and integrated approach, aimed at innovation leaders, and consisting of three modules (convert ideas into opportunities, create the compelling case, adopt and implement); this approach is supported by about 40 worksheets.

The approach of splitting projects into discrete steps, with regular review and decision points, is common practice. Early examples include the chemical industry since the 1940's and NASA in the 1960's. In the context of new product development Cooper has developed and popularised the stage-gate model, and has registered "Stage-Gate" as a trademark. An early article was entitled "Stage-gate systems: a new tool for managing new products" (Cooper, 1990). More recently he has noted that criticism of the system included it "being too linear, too rigid and too planned to handle more innovative or dynamic projects" (Cooper, 2016a). Responses have included developing variations of the model such as Stage-Gate Lite for product modifications and "renovations," Stage-Gate Xpress for very small changes or single customer requests and Stage-Gate-TD for technology development projects. One of the current developments is to explore using "Agile" methodology, which arose in 2001 from the need to develop software in a rapidly changing environment of markets, needs and customers. Many innovations start with a frustration and in this case Agile was stimulated by frustration with linear processes which were perceived as bureaucratic and ineffective. As described in an article "Embracing agile" (Rigby, Sutherland and Takeuchi, 2016), the approach focuses on people over processes, response to change rather than following a plan, working prototypes over excessive documentation, and customer collaboration over rigid contracts.

Cooper has described the first steps being taken by companies to apply the Agile approach to the development of physical rather than software products. While this author thinks that "this new Agile-Stage-Gate hybrid approach promises to be the most significant change to our thinking about how new-product development should be done since the introduction of today's popular gating systems thirty years ago" (Cooper, 2016b), he has also noted a problem: the Agile methodology requires dedicated project team members, working on only one project, and located at the same place. This is not how many innovation teams work, and so implementing Agile requires changes in how people work. It was also noted that one of the challenges is that in focusing on delivery of short term goals (sprints) the focus on the overall goal was neglected (Cooper, 2016b).

Subsequent articles (Cooper 2017a, b) have addressed other challenges, which include finding the right resources for this approach, dealing with changing product definitions, reconciling the short-term planning of Agile with the longer-term planning for a manufacturing product development project, and scaling up to larger more complex projects.

An approach to managing the innovation process in technology organizations is the use of the Technology Readiness Levels (TRLs). The TRL scale was first developed at NASA, the American National Aeronautics and Space Administration, in 1974, with the aim of establishing how far a technology was from deployment in space. Other organizations have adopted, and adapted, TRL scales. One major current application is within Horizon 2020, a European Union Research and Innovation program with about €80 billion of funding available over 7 years, from 2014 to 2020. Ten TRLs are defined in this program as follows (Schild, 2014):

TRL 0: Idea. Unproven concept, no testing has been performed.

TRL 1: Basic research. Principles postulated and observed but no experimental proof available.

TRL 2: Technology formulation. Concept and application have been formulated.

TRL 3: Applied research. First laboratory tests completed; proof of concept.

TRL 4: Small scale prototype built in a laboratory environment ("ugly prototype").

TRL 5: Large scale prototype tested in intended environment.

TRL 6: Prototype system tested in intended environment close to expected performance.

TRL 7: Demonstration system operating in operational environment at pre-commercial scale.

TRL 8: First of a kind commercial system. Manufacturing issues solved.

TRL 9: Full commercial application, technology available for consumers.

Two reviews have described TRLs. One is provided by EARTO, the European Association of Research and Technology Organizations. EARTO represents about 350 research and technology organizations, some of whom themselves have many members. For example, in Germany the Fraunhofer-Gesellschaft, one of the leading bodies for applied research in Europe, is a member of EARTO and itself made up of 69 institutes. The EARTO review of technology readiness levels (EARTO, 2014) covers background to TRLs, and their use in different organizations. A number of limitations were noted:

(i) The lack of attention to setbacks in technology maturity. For example, problems encountered at a later stage (higher TRL) can return the program to a lower TRL until the problem is solved.
(ii) Single technology maturity approach. The TRL approach deals with a development of a single technology and is more complicated when integration of different technologies is required.
(iii) Focus on product development, rather than manufacturability, commercialization and organizational changes. This has been recognized in the Horizon 2020 program which calls for business plans to be provided at higher TRLs.
(iv) Context specificity of TRL scales. The scale needs to be adapted for use by specific organizations.

The EARTO review also covers the limitations of a simple, linear approach, the problem of commercialization of research results known as the valley of death (discussed below), and the role of research technology organizations.

A second review discusses the evolution of the TRL scale from its original use in NASA to it becoming a policy tool in the EU (Héder, 2017). The article argues "that it has never been established whether the originally space and weapons technology-specific TRL scales can be used fruitfully in all areas of innovation," and, as noted in the EARTO article, that the scale must be adapted for different contexts. It was stated that "in the absence of discipline-specific guides, TRLs will predictably become a source of confusion and subject of abuse in efforts to obtain EU funding."

It remains to be seen whether the prediction in the preceding sentence turns out to be correct. In any case the scope and scale of the H2020 program means that it should provide useful evaluations of the application of TRLs.

Some organizations may prefer to adopt a complete system for innovation while others may prefer to develop their own, or to use a combination of external advice and own experience. Some extensive, company-wide systems require considerable experience to fully appreciate, and considerable authority to implement. In all cases the system should be flexible, and capable of adaptation to new circumstances, ranging from improvements to software for idea management up to major reorganizations. It is clear from the above discussion that processes for managing and stimulating innovation continue to evolve, and are themselves subject to innovation.

2.4 A PROCESS FOR EARLY STAGE INNOVATION

The focus in this book is on the early stages of innovation, depicted in Fig. 2.2, and on bearing in mind the needs of young (technical) staff in a first job. Early stage innovation is where this group is most likely to make a contribution. Many new staff have ideas about what could be done differently, and an understanding of what innovation is (previous chapter) and how to define and develop a good idea (this chapter) will improve the chance of their ideas being adopted, and thereby reduce the chance of frustration.

In the following sections, a basic process for managing and stimulating early stage innovation is described. The process for early stage innovation described in this chapter starts with a blank sheet of paper, or an initial idea, and finishes with a description of an opportunity in which the idea has been initially tested and proof-of-concept established. Proof-of-Concept is widely recognized as a stage between an idea, or promising research result, and an innovation; for example the European Research Council offers additional funding, for holders of ERC grants, to establish Proof-of-Concept (European Research Council, 2017).

The overall approach can be described as a stage-gated funnel, in combination with a portfolio of opportunities. The funnel reflects the concept of starting with a lot of small ideas, developing them, and filtering out those which fail to meet sets of criteria (the stage gates). This is illustrated in Fig. 2.3. The portfolio refers to a set of opportunities which are balanced, according to the needs, across long term and short term, impact if successful, and chance of success.

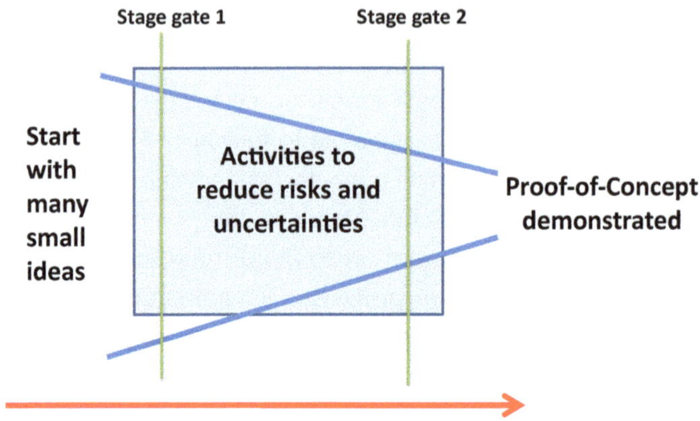

Fig. 2.3. The stage-gated funnel.

It is widely believed that it is valuable to start with a lot of ideas, and then select, rather than attempting to generate a single, very good opportunity at the outset. The comment that is sometimes made "we have too many ideas" may reflect a lack of clarity on what constitutes a good idea. In a recent article entitled "How to build a culture of originality." Grant emphasizes the value of producing many ideas, and also refers to work that has shown that the volume of ideas produced leads to improved quality: in experiments conducted by psychologists Lucas and Nordgren it was found that the first 20 ideas generated were significantly less original than the subsequent 15 (Grant, 2016).

The approach suggested in this chapter is to split the process of early stage innovation into six steps:

- Defining the scope for innovation
- Selecting inputs for idea generation
- Idea generation
- Obtaining resources
- Use of stage gates
- Establishing Proof-of-Concept

These are now discussed in turn.

Defining the scope for innovation

It is important at the outset of the innovation journey to have some idea of what an organization will accept as an innovation: some ideas will clearly be relevant, others will be clearly irrelevant, and there will be grey areas in between. For example, the management of a large energy company will in principle be prepared to consider all ideas related to energy, but will be less enchanted by proposals to manufacture chocolate. As a grey area, water management is a significant factor in many operations, but may or may not be attractive as a separate business unit providing technologies and services to other organizations.

The sense of what is in scope or out of scope for an organization will change, depending on how well it is doing, the willingness to take risks, and the resources available. It is therefore valuable to regularly revisit the boundaries and to establish if previously held assumptions and perceptions are still valid. One situation to be avoided is that where staff waits for some direction from management on acceptable scope, and management waits for new ideas to be proposed. A solution is to create a forum where ideas can be presented for a preliminary management response. Such idea presentations should cover the basic questions of what is the idea?, what is the new element?, why should we do it?, what might it deliver?, and what resources are required? Ideas can be presented in idea fairs, as pitches, or as posters, and can gain direct support, or be rejected, or fall into the category of "maybe, but not now."

While proposers of ideas are primarily interested in these reactions, the bigger picture is that a dialogue is being created which shapes, and shares, the boundaries of what is in scope. Effective communication on such sessions will ensure that a common awareness of acceptable scope develops. The issue of scope also refers to different time-scales, from those covered by a current strategy, to those involved in innovative proposals which may take longer to achieve significant impact. It is thus desirable that management is prepared to discuss efforts which fall outside the current strategy either in terms of content or time-scale.

The discussion on scope can be held at multiple levels: from individual projects up to the whole organization, and the same principles apply: explore the grey area between what is thought of as in-scope and what is considered out-of-scope, on the basis of initial proposals, and through a climate in which this is seen as a good use of time for all parties concerned.

Selecting inputs for idea generation

Having assessed the scope, and before moving into idea generation, it is useful to consider what inputs can be used. In scientific and technical areas, an organization's own staff will usually have at least some relevant expertise. This can be augmented with literature reviews to assess the state of the art, and by bringing in external experts to contribute. The latter can be arranged on a non-confidential basis, in which the primary aim is to listen to experts, or under a secrecy agreement if active dialogue pertaining to new opportunities is required.

Customers can be a valuable source of information, and negative comments can be seen as a potential opportunity. Customers are not necessarily right though: recall the comments (Chapter 1) attributed to Henry Ford: "If I had asked my customers what they want they would have said: faster horses" and made by Steve Jobs: "A lot of times, people don't know what they want until you show it to them." As discussed in Chapter 1, recognizing unarticulated needs can be a powerful source of innovation.

In large organizations with exploratory R and D departments the primary customers will often be internal and there will, or should be, regular discussions on strategy and goals.

Another input is obtained by reviewing competitor activities, and perhaps mapping them against one's own projects to assess where there is overlap and where there is a unique space. In scientific and technical areas, analysis of the patent literature can be useful, and there are increasingly sophisticated tools to analyze and visualize the data available (Stalbauer and Drexler, 2014).

Looking for trends and changes can generate useful inputs. There is a steady stream of books focusing on what the future might be (Franklin and Andrews, 2012; van Santen, Khoe and Vermeer, 2010; Watson, 2010) and although there are notable failures in predictions about the future, many serve to challenge and extend boundaries of thinking. Statements such as "the Americans have need of the telephone, but we do not. We have plenty of messenger boys" by Sir William Preece, Chief Engineer, British

Post Office, in 1878 (Cellan-Jones, 2012) may or may not have sounded plausible at the time, but can be used to stimulate ideas — under what conditions would a telephone be better than a messenger boy, and how might such conditions develop? While Preece's statement seems quaint nowadays, it is less easy to see what the future will hold for the plethora of communication devices and channels currently available.

Awareness of changes — technical, social, commercial, regulatory — can provide valuable inputs to idea generation. Changes can be fast or slow, and the former are easier to perceive and act on. This is illustrated by the description (which may not be true, but which is still a useful image) of a frog dropped into a pan of hot water and immediately jumping out, compared to a frog being placed in cold water which is slowly heated until it is no longer able to jump. An example is climate change: a rapid series of extreme weather events, such as major hurricanes, will likely produce more action than reports of gradually rising sea levels.

A particular subset of changes is covered by the concept of "weak signals" — early evidence of upcoming trends and changes. An article "Planned opportunism: using weak signals to spur innovation" (Govindarjan, 2016) illustrates the importance of identifying and acting on changes which can affect a company. One example discussed is the early perception by the games company Hasbro of disruptive technologies, new distribution channels, fundamental customer shifts, non-traditional competitors and globalization, and the subsequent strengthening of their competitive position based on knowledge of these factors.

Scenarios are a useful tool to bring together many possible changes into a coherent picture. Like the word innovation, the word scenarios is widely used, and in diverse areas, including business strategy or film-making. In the context of business, scenarios are stories about the future which are plausible, relevant and challenging; they are not predictions or forecasts (Scwharz, 1998; Schwarz, 2011; van der Heijden, 2005; Ramirez and Wilkinson, 2016). Scenarios enable different futures to be envisioned, and offer a way to explore deeply-held assumptions about how the future will turn out. One of the leading texts (van der Heijden, 2005) on scenarios is sub-titled "The art of strategic conversation" in order to emphasize the role scenarios can play in structuring and stimulating conversations about the future, and the associated strategic options.

Verloop has discussed scenarios in relation to innovation (Verloop, 2007) and points out that scenarios have a particular role to play in radical innovations, which depend on new markets, products or services, since scenarios provide a basis for discussing how different futures could unfold. A major new initiative which does well in several different possible futures is more robust than one which relies on a specific set of circumstances. It must be remembered that scenarios are not predictions: what actually happens may not correspond to any one scenario. However, through developing different scenarios, and thinking through the implications, an organization is better prepared — and less likely to be surprised — by what does transpire.

Scenarios have been linked to innovation through the concept of domains (Ramirez, Roodhart and Manders, 2011). Domains are described in this work as sets of strategic options, and associated projects, which act as stepping stones between possible futures, derived from scenarios, and existing activities. The article explores this concept with reference to practice in Shell, and also discusses how actor-network theory can contribute to understanding of this practice.

Examples of scenarios include those produced by the World Economic Forum, which cover possible futures for regions such as Europe, China or the Russian Federation, as well as sector-specific scenarios, such as those on the Future of Oil Demand, and on the Metals and Mining Industry. Scenarios on many other topics, produced by a variety of organizations, can be readily found.

Generating ideas

A common image of idea generation is that of the flash of inspiration, illustrated by the story of the Greek scholar Archimedes jumping out his bath and exclaiming "Eureka!" ("I have found it"). This was due to his observation of the rise in level of the bath water when he got in to the bath, and the realization that the volume of the water displaced was equal to the volume of the submerged part of his body. In modern times — post Edison — the image of a light bulb suddenly lighting up in or above someone's head is a widespread way of indicating inspiration. Although ideas can appear to arise spontaneously in one's mind, it is likely that this is preceded, consciously or unconsciously, by awareness of possibilities, problems and opportunities.

The 19th century chemist Kekulé made a major advance in organic chemistry by proposing that the six-carbon benzene molecule had a cyclic structure, with the carbon atoms arranged in a hexagon; he said the idea for this came from a day-dream involving a snake holding its tail in its mouth. This image was itself known as a symbol, Ouroboros, in Ancient Egypt. In a similar example, the inventor Elias Howe made several contributions to the invention of a sewing machine and one of these concerned the issue of where to place the eye of the needle. He reputedly thought of the solution in a dream, where he dreamt he was to be executed for failing to complete the invention in time. The executioners were warriors with spears which had holes near the point of the spear, and this became the effective design. While many would recognize the beneficial stimulus of deadlines and working under pressure, it must be hoped that it is not necessary to apply fear of death to achieve innovation.

The role of dreams in leading to ideas is discussed under "Serendipity," one of seven patterns of innovation identified in the book, *Where Good Ideas Come From*, which covers many innovations over many centuries (Johnson, 2010) (see also Chapter 1 for serendipity in relation to research).

There is no doubt that many good ideas start with individuals. Scientists and engineers engaged in research are focused on interpreting data, formulating hypotheses, designing experiments and absorbing information from other research. Much of this has as a strong individual component, although of course constant refinement and extension comes from dialogue with others, informally or through presentations and publications.

In addition to individual creativity, increasing attention has been paid to achieving more and better ideas through using groups, with the aim of combining insights, ideas, expertise and experience from different individuals. An early example came from the work of Alex Osborn, an advertising executive, who began developing group approaches to creativity in 1939, and who popularized the term brainstorming. The book *When Sparks Fly — Igniting Creativity in Groups* combined insights from the authors' backgrounds in psychology and business (Leonard and Swap, 1999).

These approaches emphasize the need for a divergent phase in which many ideas are created by a group, with judgment postponed to a later stage. Leonard and Swap advocate application of an incubation phase following the divergent phase, and then a convergent phase to arrive at the best ideas or options.

So stimulating idea generation requires mechanisms to be available for both individuals and groups. The classic wooden box for individual employees to submit suggestions has made way for increasingly sophisticated software systems, either developed in house, or purchased. In addition there will be ways as part of normal business processes for individuals to propose ideas as extensions of existing programs, or as new programs. All ideas submitted should get a clear response within a reasonable period of time in order to maintain the motivation of submitters and to keep submissions flowing. A database of ideas submitted should be established, in order to regularly review ideas which are initially rejected: such ideas can become viable either because the context has changed, or because it is worthwhile to add them to other initiatives.

With respect to using groups to generate ideas, these can be informal discussion groups, tasked to meet regularly and generate options, or structured workshops which are carefully designed and facilitated. Such workshops can be expensive if they involve significant amounts of preparation time, participant time and travel, and should only be undertaken with proper preparation. Such preparation should include a crystal-clear formulation of what the workshop intends to achieve, establishing that resources are available to follow-up on ideas generated, and plans to document the workshop output and to communicate the follow-up.

More detailed steps in designing a workshop include selecting internal or external participants, devising a process for idea generation (for example ways to get from an agreed starting point to an agreed goal), and criteria for ranking ideas generated (for example: what is the impact of this idea if it is successful? what effort is required to implement this idea? what is the chance of success?).

There are many tools and techniques available for idea generation. Different teams or organizations will have their own preferences and probably the best approach is to experiment with different techniques on a small scale. If this is driven by some enthusiastic participants it is likely to be more successful than a system which is imposed. While it is not a problem for different idea generation techniques to be used, the output of such sessions should have the potential to be shared, which may not be the case when using different proprietary systems for idea generation and documentation. In general ideas should be freely shared as much as possible, since this allows the opportunity for cross-fertilization, but in some cases it may be appropriate to restrict access to topics which are particularly sensitive.

Generation of ideas can be stimulated by exploring opportunities, needs, and problems (including reframing), as well as searching for, combining, and adapting technologies from other fields. These approaches are discussed in Chapter 1, together with R and D. Some additional points are:

Combining skills

Reviewing the potential for unusual combinations of skills present in a group can lead to new possibilities, and can be further stimulated by including participants from outside the organization. This approach can also be used in terms of looking for productive combinations between organizational units (for example teams or departments), or between technologies.

Hurdles and jumps

One relatively easy systematic approach, suitable for use in team discussions or workshops, is to define both the starting point and the goals, using formulations which are crisp, clear and agreed, and then work on defining barriers which need to be overcome, and ways to cross the barriers. This process can be depicted as a staircase of "hurdles" (the barriers) and "jumps" (the solutions), which links the starting point and the goal.

TRIZ

An approach to idea generation in the context of problem solving is TRIZ, an acronym derived from the Russian phrase "theory of the resolution of invention-related tasks." This approach, which started in Russia in 1946, and which became more widely available after the end of the cold war, is derived from analysis of trends in the global patent literature. An article on innovation in Samsung has stated that TRIZ has become "the bedrock of innovation" at that company (Shaughnessy, 2013). Information on TRIZ can be found on the web sites of the Altshuller Institute for TRIZ studies, and of the European TRIZ association.

In applying these approaches to idea generation it is helpful to first do a quick scan of opportunities, followed by more detailed review, rather than getting bogged down in analysis of the first option which arises.

Developing the idea: Resources

All organizations have to make choices with respect to allocation of resources such as staff and money, and in the case of technical centers, of equipment as well. If things are going well, it may be possible to expand budgets to accommodate innovation efforts. If, however, cost-cutting is dominant, then budgets for new initiatives may have to be re-allocated from other projects, and this can be a source of tension if the reason for the reallocation is not explained convincingly.

An organization may be limited in resources to pursue innovative projects either because the available resources are committed, or because developing the idea requires expertise not available internally. In both cases it makes sense to look to the outside world, and assess whether there are opportunities for collaboration or contracted projects. Establishing such types of cooperation can take significant amounts of time and energy, and it should be ensured that the benefits outweigh the effort of establishing an external program. Use of external resources, while long established, has received much attention in the last 15 years through the emphasis on Open Innovation, and this topic is explored in more detail in Chapter 3.

Stage gates

The premise behind the stage-gated funnel is that producing innovations requires starting with many ideas, developing them initially on a small scale, and filtering out those which are less promising. The use of two filters, or stage gates, has proved useful in early stage innovation, in which the first filter is applied on mainly qualitative criteria, and the second filter includes more quantitative data, for example on technical or market results, as well as estimates of costs and revenues. It may be appropriate to have more than two stage gates, and major projects, for example in engineering or infrastructure, require many more.

In applying the filters, it is important to maintain focus on the overall purpose, which is to establish Proof-of-Concept, discussed below. A set of questions is given there which typically need answering to establish Proof-of-Concept, after which the opportunity can be developed further. These questions can be prioritized according to the effort required to provide answers, the resources available, and the perceived importance of the answer in determining further work. Questions can be grouped into sets, with each set forming one filter, or stage gate. Some answers will be sufficient to terminate further development of the opportunity, for example if the potential revenues are insufficiently interesting. Other answers may require further work. The overall result of generating answers is to weed out ideas which are not attractive. The funnel starts with many potential opportunities, and ends with a few which meet the requirements of proof-of-concept, and which are then considered ripe for further development.

Making and responding to proposals

A call for new ideas means that there also has to be an adequate way to deal with the ideas produced. In order to keep proponents motivated to keep producing ideas it is important that the response is fast, that the grounds on which a proposal is supported or rejected are clear, and that the attitude of those judging the proposal is one of constructive criticism.

A serious commitment to providing a fast response to proposals is necessary because it is all too easy for a proposal, which at first sight does not look promising, to end up somewhere near the bottom of a "to do" pile. Proposals which should clearly be supported, or those for which there is no doubt that they should not, are easy to deal with; many, however, are in an indeterminate category. To avoid subjective decisions it is better if proposals are judged by a small group than by individuals. Acceptance of the decision on a proposal is increased if those making the decision have credibility with those submitting the proposal.

With respect to achieving clear, and shared, understanding of the decision to support or reject a proposal the criteria should be well-formulated and actively communicated. Active communication is necessary since some people only give a cursory glance at guidelines for proposals (innovative or other) and are then disappointed to receive the response that the proposal does not merit consideration in its present form.

Those making proposals are therefore encouraged to take the guidelines seriously. Speaking for lot longer than the allotted time in a pitch, submitting written proposals which are much too long or too short, or which do not address the required criteria, does not improve the mood of those asked to judge proposals.

Those judging proposals should aim to provide constructive criticism. Disregard of guidelines may be difficult to deal with, especially in the context of many other pressing commitments, but sometimes guidelines are not as well known as is assumed. Furthermore, some people may see exposing the weaknesses in a proposal as a robust, legitimate and necessary activity. As one senior executive commented: "Our approach is to keep shooting (questions) at the proposal, and if the proposer is still standing at the end, there may — possibly — be something in it." This approach does not work well in all cultures.

Establishing Proof-of-Concept

Proof-of-Concept, also known as Proof-of-Principle, refers to a stage in the development of an idea. At this point the idea is considered feasible. There is not a rigid template for Proof-of-Concept, since different industries, and different organizations, will have their own requirements, but these will very likely include answers to the following questions. The discussions generated by the questions will also be valuable in sharing and aligning perceptions. The output from early stage innovation is thus a document which addresses the questions, summarizes inputs received, establishes Proof-of-Concept to the satisfaction of the organization concerned, and thereby forms the basis for a decision on further development.

1. What is the idea?

The description of the idea should be concise, and clear to both experts and non-experts. This is usually more work than initially realized, but getting a crisp formulation helps to clarify strengths and weaknesses in the proposal. Acronyms, technical or otherwise, should be avoided as much as possible; otherwise the proposal can end up looking like gibberish to the uninitiated.

It is useful to put the idea into a context of a bigger picture of changes and trends, and to start with these before proceeding to the idea, rather than the other way around. It is also helpful to include in the description of the idea the need which it is addressing, and as discussed in Chapter 1 this can be a clearly recognized need, or an unarticulated one. If the need is known, then the question arises as to whether it is already being met or not, and this is addressed below in relation to competitors.

2. What is the new element?

In order to be consistent with the definition given in Chapter 1 — innovation is the successful exploitation of new ideas — it should be clear what the new element is. This might be a new technical advance, or a new market. As discussed in Chapter 1, it is usually not advisable to attempt to tackle both at the same time. Description of the new element should include answers to the questions of "How new?" and "New to whom?" (see Chapter 1).

Note that "new" also covers an old idea which was not viable in the past but is now, due to changing (new) circumstances. Such changes include a technical idea which has become cheaper to bring into practice, combinations with other technologies which were not available in the past, new markets, and changes in regulations. The reaction of "that won't work, we've tried it unsuccessfully in the past," should always be countered with an exploration of changes in the environment which would now make the idea viable. Taking this a step further, it can be worthwhile to maintain a database of proposals and regularly revisit proposals which were previously rejected to see if any now merit support due to changes in factors such as cost, benefits, and regulations.

3. What are the benefits and costs?

At an early stage, it is not worth trying to do a detailed financial analysis of the benefits, and development costs, of an innovative idea. There are too many uncertainties involved to make the result credible. It is, however, necessary to generate an order-of-magnitude estimate of benefits from a first successful deployment of an idea, and of the costs which would be involved, as well as an assessment of the opportunities for scale-up or further deployment. As the opportunity is developed the accuracy of the estimates should be increased.

The value of the opportunity is not necessarily just financial: there may be reputational factors, or the opportunity may provide access to partners who are valuable in other projects.

4. How does the opportunity fit the strategy?

Questions to be addressed here include: does the opportunity fit the existing strategy, require an extension of strategy, or require a new strategic direction to be established? Incremental innovations will fall under an existing strategy whereas radical innovations may require a new strategic direction to be established.

A related aspect is the opportunity cost: does developing the proposed innovation mean that other projects will not be able to go ahead, and is there clarity on the reasons for the choice?

5. Who are the competitors?

If the need is known, there may be competitors with an offering (product or service) which is meeting the need. It is then necessary to be clear on the specific features of the proposed innovation which will be more attractive to customers, and to understand why they would, or would not, switch to the new offering.

6. How can the opportunity be protected?

Assuming that a plausible case can be made to embark on an innovative project, the question of protecting the opportunity needs to be addressed. In other words, if the project was successful, what would prevent others from copying the idea? The answer might lie in patenting, but filing and maintaining patents on a global basis is expensive. Other approaches include being first to market, or relying on expertise — exactly how to do something — which is not readily patentable, but also not readily reproduced, or by locking-in partners.

7. What key unknowns have been identified and addressed?

The uncertainties associated with a new idea should be thought about at the outset. Does a technical idea require some additional experimentation to support it? Or is the technical case strong, but the market uncertain? Reviewing and prioritizing the unknowns generates valuable discussion, and identifying the key unknown leads to the first steps. More technical data to support the case can be obtained from literature reviews, own experiments, or contract research. Assessing market opportunities can be done by talking to likely customers to gauge their interest. Here the conversation should focus not just on the opportunity, but on gaining a deeper understanding of the need that the proposal would meet. In some cases it might be appropriate to commission a small market survey.

In taking these first steps the focus must be on whether the information obtained has significantly reduced the risk associated with the project.

8. What are the next steps?

Work on identifying and addressing key unknowns may generate satisfactory answers, or may require further work, or may lead to new questions. These should be included in an overall plan which covers all the information gained from addressing the previous questions, and describes the next steps required, along with the costs and timing involved.

2.5 AFTER PROOF-OF-CONCEPT

Satisfactory answers to the above questions establish Proof-of-Concept, and the opportunity can then be compared to others. Funds can only be spent once, and so it is possible that a good opportunity has to take second place to an even better one. In this case development of the opportunity can be postponed until funding is available, or it can be spun out; this latter option forms part of Open Innovation, discussed in Chapter 3.

In small organizations, or with incremental innovations, it may be possible for one person to lead the innovation project from idea to market impact. In many other cases a handover will be required, between individuals or from one unit to another, for example from an exploratory research department to a development department to a business unit. It may also be true that the proponent of the original idea is not the best person to champion the project in later stages. For example, academics who have an idea for a business based on their research, and who approach venture capitalists for funding, are sometimes told "the good news is that we are going to fund development of this opportunity, the bad news is that you will not be running it." In this case there is a clear commitment to develop the opportunity, but this is not always the case for project hand-overs within an organization. Those on the receiving end may consider that the opportunity merits less of their attention than needed by their existing commitments, and this is of course frustrating to the proponents of the opportunity.

This problem of the handover is frequently referred to as the "Valley of Death": the project dies between the people or department where it originated, and commercialization; it can be seen as the gap between invention and innovation. This problem is considered to be one of the major issues in making innovation work and is a concern to governments, universities and companies. The cases studies presented in Chapter 5, however, show various successful approaches to crossing the valley of death.

"Surviving the Valley of Death" is the title of a special issue of the journal Technovation, with publication planned for 2019. The call for papers (Technovation, 2017) suggests the following topics, which indicate the different facets involved:

- Defining the boundaries of the Valley of Death and suggested processes to cross it.
- The tensions between invention and innovation within organizations
- Strategies and approaches to integrate technological development and product development
- The role of champions in transitioning from invention to innovation
- The role of collaboration between industry, academia and government agencies
- The role of various financing instruments to support technology development
- The role of technology policy and government support for different stages of innovation (basic research, intermediate feasibility analysis, product development)

- The impact of industry context on the characteristics of the Valley of Death
- The Valley of Death in startup firms versus established companies, and the instruments used to coordinate technology development and product development

Within an organization, crossing the valley of death requires communication and trust to be established between the proposers of a potential innovation, and those who should develop it further. This can be made more difficult because of the different cultures involved. There are different ways of thinking and working on each side of the valley, from exploratory and divergent on the one side, to focused and convergent on the other. Those at the front end of the process should remember that the hard and focused work of generating profits is necessary to make the front end possible; those at the commercial sharp end should remember that the profitable business started as an untested idea.

One way to overcome this is to have someone with responsibility, and sufficient authority, to bridge the gap and ensure smooth transition. In a short essay on the future of innovation (part of a collection of essays on this topic), Hissel suggested that the future of innovation lies with Innovation Chain Masters, who oversee the whole process (Hissel, 2009).

Markham and Mugge take a broader view of the Valley of Death, and use the term to describe the space between ideas and markets; the three modules referred to above (convert ideas into opportunities, create the compelling case, adopt and implement) aim to support the safe journey through the valley (Markham and Mugge, 2015).

The review of Technology Readiness Levels by EARTO, noted above, stated that the "Valley of Death" is ostensibly the reason why the TRL scale has been adopted by the European Commission (EARTO, 2014). The evaluation of the current Horizon 2020 program, which runs from 2014 to 2020, will provide evidence on how effective TLRs prove to be in bridging the Valley of Death, as well as their effectiveness more generally.

Although developing the business case is outside of the scope of this book (Fig. 2.2) it is worth mentioning a recent approach to business model generation which has become widely used (Osterwalder, Pigneur and colleagues, 2010). A business model describes how an organization creates and captures value. Their approach is based on nine headings ("building blocks"), which are used to capture all the relevant ideas and insights. These building blocks are:

1. Customer segments
2. Value propositions
3. Channels
4. Customer relationships
5. Revenue streams
6. Key resources
7. Key activities
8. Key partnerships
9. Cost structure

These nine building blocks can be displayed on a single sheet, or poster, known as the Business Model Canvas, which enables groups to discuss the individual building blocks and how they fit together.

Other resources which may be useful to those wishing to explore what comes after Proof-of-Concept include a current, annual series of conferences called the "Back End of Innovation," which focus on the issues of taking a new idea to market. A short book entitled, *Zero to One: Notes on Startups or How to Build the Future* (Thiel, 2014) is based on a course given by the author at Stanford University in 2012.

2.6 BALANCING, ACCELERATING, AND MEASURING SUCCESS

So far the need for a process to manage and stimulate early stage innovation has been discussed, and six steps described. This chapter concludes with three further factors of importance: the need to have a portfolio approach, the need to accelerate the flow through the funnel, and the need to measure success.

The portfolio approach

The portfolio approach reflects a set of options which should be balanced across chance of success, impact if successful, and resources required. It is useful to have some ideas in the category of high chance of success, limited need for resources, and small, but not negligible, impact if successful, since such projects help to generate credibility for the innovation efforts. The willingness to accept riskier projects, with potentially a bigger impact, will then increase. In some circumstances, with a very compelling opportunity, it might be considered appropriate to bet all the available resources on this single option. In this case it would be wise to pay significant attention to the possibility of the whole organization becoming insufficiently critical, in other words that group-think starts to play a role, and to put mechanisms in place to ensure assumptions are rigorously reviewed on a regular basis.

Accelerating the flow through the funnel

The concept of the stage-gated funnel is easy to understand, and looks neat on paper. In practice the flow through the funnel is not as smooth or as fast as is wished: priorities change, for example immediate business needs take priority over early stage innovation, or individuals move on to other positions. An important role in managing innovation is to ensure a smooth flow through the funnel. Two mechanisms can help: assigning project coaches, who can act as a second pair of eyes; here a relationship of trust which will facilitate constructive criticism is important. Another approach, as noted above, is the "innovation chain master" who oversees the whole process and identifies bottlenecks. Clearly such a person should have the contacts, credibility and authority to enable bottlenecks to be overcome.

With respect to the number of ideas coming into the funnel, there are various possibilities:

- More good ideas than there are resources to develop them. In this situation a case can be made for additional resources, or more rigorous selection can be applied, including the option of postponing (but not forgetting) development of some ideas to a later date.
- Many ideas, but of mixed quality. Here the solution is to make sure the criteria for idea selection are clearly formulated and better communicated.
- Too few ideas. This requires re-examining the commitment of the organization to innovation. Paying lip-service to innovation, but

allowing short term business needs to always take priority, may require (senior) management intervention to ensure that a healthy balance is maintained.

Measuring success

Any commitment of resources by an organization to innovation will sooner or later lead to the question of whether or not this was money well spent, which leads into the discussion of how to measure success. This can refer to both individual projects, as well as the investment of time, money and people in an innovation management system.

While measuring success is important, it should be noted that not all innovations have an impact that can be readily quantified. For example, the Nobel prize-winning economist Joseph Stiglitz has pointed out the difficulties of measuring the impact of ATM's, on-line grocery shopping, or improved medical procedures (Stiglitz, 2014). The first two involve putting a value on convenience, and in the case of ATM's taking into account job losses amongst bank staff; quantifying the impact of improved medical procedures requires a value to be assigned to reduced pain, discomfort, or disability, or to the extension of life.

Some extensive schemes for measuring innovation have been developed. The European Commission provides a range of tools for assessing innovation performance within the EU; one tool, the European Innovation Scoreboard 2017, compares countries' innovation performance on the basis of 27 indicators (European Innovation Scoreboard, 2017). Companies will usually have simpler assessments, and focus on the revenues derived from new products or services.

With respect to early stage innovation, from idea to Proof-of-Concept, there is not a single magic metric but assessment of the following factors, individually or in combinations, will be valuable.

Number of new ideas

The number of ideas generated or proposed in a year gives some idea of the enthusiasm and creativity within the organization. In the preceding section options were described for dealing with too many or too few ideas.

Number of ideas funded

This provides a judgment on the quality of the ideas. If only a small proportion is funded, it may be necessary to put more effort into making the funding criteria known. If a high proportion is funded it may be appropriate to encourage more ambitious proposals.

Estimated value

As noted in the discussion of the Proof-of-Concept, an order-of-magnitude assessment of the value of the proposition should be made, as well as of any impact which is not directly financial, such as reputation or access to new partners. The estimated value should be reviewed regularly, not just because of the reduction in uncertainty, but also because assumptions may no longer be valid, for example if the context has changed.

As ideas progress through the funnel to handover and successful launch, the uncertainty about the value of the idea decreases. However, usually many different contributions are necessary to take an idea through to commercial impact. If this is achieved, assigning a value to one idea which forms part of the whole is not likely to be easy or credible.

Customer satisfaction

Dialogue with customers, internally or externally, will indicate levels of satisfaction and generate suggestions for improvements. Customers within an organization for ideas which have successfully demonstrated proof-of-concept include those who are taking the idea further as well as those who funded the early stage innovation. Active dialogue is necessary to reach a common, shared, perception of what has been achieved, and to avoid both over-selling and under-appreciation.

Look-backs

Annual reviews of individual or departmental performance are common. It can also be valuable to undertake a review of ideas generated and developed over a longer period, with the aim of learning which assumptions were (in)valid, and why, and whether past ideas which were rejected are now viable because of changes in the context. This forms part of the importance of learning (Chapter 4).

Termination of projects is not a failure

Finally, it should be noted that termination of projects should not be seen as a failure. Early termination represents a saving of expenditure on unnecessary further development. Clarifying, agreeing on, documenting and sharing the reasons for project termination is valuable, for use in look-backs as described above, but also to ensure that those who committed much time and energy to the projects are motivated to try again. However, as projects start to die, those involved are keen to move on to something new, and the opportunity for learning can be lost (Chapter 4).

3 Open Innovation

3.1 INTRODUCTION

The topic of open innovation has received considerable attention since 2003, when an influential book on the subject appeared (Chesbrough, 2003a). Open innovation is now an important component of the innovation capability, and is also the subject of much academic research. It is a mechanism through which an organization engages systematically and strategically with the world outside its boundaries in order to enhance the benefits of collaboration. This interaction can take the form of knowledge flows in both directions, from the outside world into the company, as well as the other way around. In addition to knowledge flows, the innovation which results can be introduced in to the organization's own markets, or in to other markets. These possibilities are shown in Fig. 3.1. This is sometimes called the porous funnel, as an extension of the stage-gated funnel (Fig. 2.3): in open innovation the walls of the funnel become porous, enabling two-way interaction with the outside world to take place, along the whole trajectory of the funnel.

Much of the initial focus of interest in open innovation was on large companies bringing in new (technical) information from the outside world, but the scope of open innovation continues to expand, with interest in application of open innovation to small and medium-sized enterprises (SMEs), to service industries, to business models, to developing countries, as well as ongoing application of open innovation in different sectors, and across different phases of the innovation journey. It is thus highly likely that open innovation will remain significant to both practitioners concerned

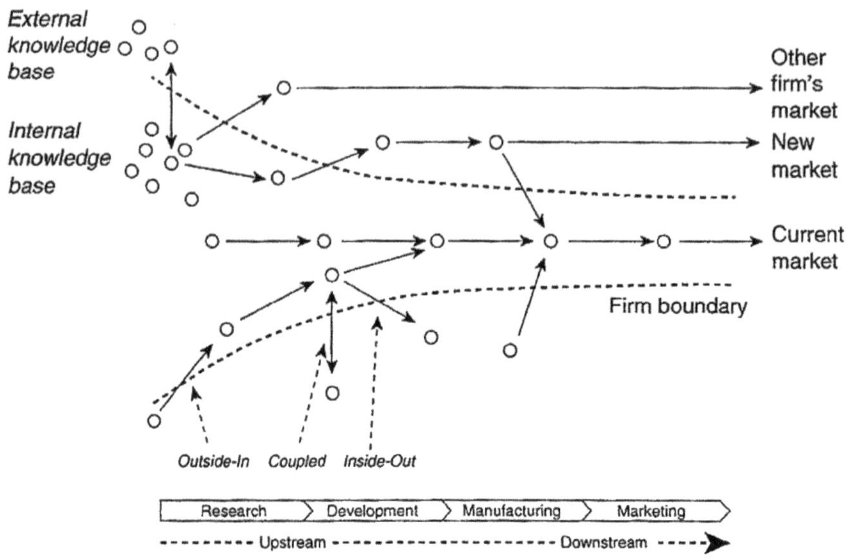

Fig. 3.1. The open innovation model. (From "New Frontiers in Open Innovation", Chesbrough, Vanhavebeke and West, 2017. Fig. 1.5, p. 18. By permission of Oxford University Press.)

with applying open innovation, as well as to researchers concerned with understanding it. Many scientists and technical staff will be involved in external collaborations, and an awareness of the many forms open innovation can take, as well as of the disadvantages, will increase the chance of using this approach effectively.

3.2 CLOSED INNOVATION

Before discussing in more detail what open innovation entails it is useful to understand what came before open innovation. Not surprisingly, this is seen as the era of closed innovation. In this period, which covers much of the 20th century, the predominant model was that of organizations producing innovations within their own boundaries, as exemplified by large industrial laboratories whose R and D activities covered the whole span from basic research through development research and technical support to commercialization and marketing; as described in Chapter 1 this does not necessarily imply that innovation proceeds along this sequence.

The era of closed innovation is illustrated by use of the phrase "EMR." This phrase, standing for Extra-Mural (beyond the walls) research, has been used to refer to external collaborations, in particular collaborations with universities by a technical organization. The phrase captures, and reinforces, the image of large industrial laboratories as citadels engaging only to a small extent with external co-operations.

Of the many examples of innovations which have arisen from large industrial laboratories, three serve to illustrate the impact of closed innovation, and the power of having many different skills, both technical and commercial, within the same organization. The first is given in Chapter 5, where the development by General Electric of a new technology for inspection of aircraft engine components is described. The other two examples deal with Nylon and Gas-to-Liquids.

Nylon

The discovery and commercialization of Nylon by DuPont has been described as an example of closed innovation (Vanhaverbeke and Chesbrough, 2017a). Here some more technical detail is added, in line with the focus of this book.

The discovery of Nylon started with Wallace Carothers joining a newly constructed DuPont laboratory in 1928, having previously been a lecturer in organic chemistry at Harvard University. Carothers started work in a department of fundamental research, on polymers, long molecules built up from simple molecules, and which occur both in nature (e.g. cellulose, DNA) as well as — nowadays — in a very wide range of synthetic molecules including rubbers and plastics (e.g. neoprene, polythene, polyurethanes). In the 1920's the nature of polymers was not well understood, and the field was met with "open contempt" (James, 1993) by the organic chemistry establishment of the day: the leading pioneer in work on polymers at the time, Herman Staudinger, was attacked by eminent scientists during a lecture to the Zurich Chemical Society in 1925, until he finally quoted Martin Luther: "Here I stand and can do no other" (James, 1993). In 1953 Staudinger received the Nobel Prize in chemistry for his work on polymers.

Thus the background to Carother's work at DuPont was not well established. Nevertheless, progress was rapid. The polyamide known as

Fiber 66, and later as Nylon, was first synthesised in 1934. This material was selected for full-scale production in 1935, and work started on building a plant to produce nylon in 1938. Production of nylon started in 1939, and a major application was nylon stockings, which became an immediate success. The first day nylon stockings were introduced across the USA, May 15, 1940, nearly 800,000 pairs were sold and seven months later, the company put 4,000,000 pairs on sale nationally.

By the start of the construction of the first plant in 1938, about 230 chemists and engineers had worked on the project. This required collaborations between different departments, and reassignment of staff. Further information is available in two booklets, "The establishment of modern polymer science by Wallace H. Carothers" (Keifer, D.H., 2000) and "The first nylon plant" (McAllister, J.F., 1995).

The speed and effectiveness of nylon discovery, development and commercialization shows what can be achieved within a single organization with both broad and deep capabilities.

Gas-to-liquids

A second, more recent, example of closed innovation concerns the development and commercialization of Gas-to-Liquids (GtL) technology by the energy company Shell. This technology converts natural gas into higher value products, including fuels, lubricants and detergent feedstocks, and provides an alternative to bringing natural gas to markets by pipeline or through liquefaction. The most recent manifestation of the Shell GtL process is in Qatar, where a plant has been producing 140,000 barrels a day of liquid products since 2012. The development of this process, within Shell, took decades, cost $18 billion-$19 billion and is covered by 3500 patents (Shell, Pearl GtL, 2017). The technology has a long history, with a related technology of Coal-to-Liquids being developed in Germany in the 1920's. Both Gas-to-Liquids and Coal-to-Liquids depend on: (i) converting the gas or coal in a high temperature partial combustion step to synthesis gas, which is a mixture of carbon monoxide and hydrogen, (ii) converting the synthesis gas over specially designed catalysts to long chain hydrocarbons and (iii) upgrading and then separating the desired components. The Shell work in research and development on GtL started in the 1970's, as a response to the rapid increased in oil prices in that decade. The work was mainly carried out within Shell, at the company's Amsterdam laboratory, which was founded in 1914. A book on the history of the Shell GtL process was published in 2014 as part of the centennial celebrations of the Amsterdam laboratory (van Helvoort, van Veen and Senden, 2014).The GtL project encompasses (to date) 40 years of research, development and commercialization, from basic research in the 1970's, through a pilot plant built in 1982, a first commercial plant in 1993, in Bintulu, Malaysia, and a world scale plant in Qatar in 2012.

The GtL development led to many innovations in different technical areas, for example burner design, catalysts, equipment scale-up and operational control and this required a laboratory with broad and deep skills. It is also essential to maintain continuity in technical expertise against changing business perspectives over such a long period. The story of GtL, like that of Nylon, demonstrates what can be achieved in closed innovation. In reviewing GtL as an innovation, four main stages were described: (i) identifying the areas in which an innovation is sought, (ii) imagining a solution or improvement, (iii) working out the concept properly, and (iv) seeing to it that the market accepts the product (van Helvoort, van Veen and Senden, 2014). While GtL is not typical of many innovations in terms of duration, these stages have clear parallels to the processes described in Chapter 2.

3.3 FROM CLOSED TO OPEN INNOVATION

The move from closed innovation to open innovation has been stimulated by a variety of factors, including the expansion of higher education around the world, giving a larger pool of talent; electronic communications which provide means of access to this talent; the rise in sources of funding such as venture capitalists; and the US 1980 legislation known as the Bayh-Dole act, which allowed organizations other than the federal government to commercialise patents generated through use of federal funding.

Against this background, Chesbrough coined the phrase "Open innovation" in two publications in 2003 (Chesbrough, H., 2003a, 2003b), and has published extensively on the topic since then, as have many others, with annual citations of 'open innovation' in Google Scholar increasing from zero in 2001 to 3500 in 2011 (Chesbrough and Bogers, 2017a). Short books (Gaule, 2006) or reviews (Huizingh, 2010; Busarovs, 2013) provide useful summaries of early work. More recent books are described at the end of this chapter.

3.4 DEFINITION AND CLASSIFICATION

As noted above, open innovation is concerned with the flow of knowledge in both directions across the boundaries between an organization and the external world. Like the word innovation, there are many descriptions and definitions of open innovation; a recent one offered by Chesbrough is that open innovation is "a distributed innovation process that relies on purposively managed knowledge flows across organizational boundaries, using pecuniary and non-pecuniary mechanisms in line with the organization's business model to guide and motivate knowledge sharing" (Chesbrough, 2017b).

A useful classification of open innovation has been made in terms of whether the knowledge flow is into the organization, or out of it (inbound or outbound), and whether this was accompanied by a financial transaction (pecuniary or non-pecuniary). This gives four cells: (Dahlander and Gann, 2010).

Here the categories are illustrated, in part with examples from Dahlander and Gann, who also provide a more extensive discussion of each category in terms of definition, advantages and disadvantages.

Revealing (outbound, non-pecuniary)

This covers the case where organizations freely share information in order to build innovation across a community of firms. One example is the iron production industry in 19th century England (discussed further below), where designs and performance of blast furnaces were shared. Another example is the publication of results not deemed commercially sensitive.

Table 3.1. Classification of open innovation

	Inbound innovation	Outbound innovation
Pecuniary	Acquiring	Selling
Non-pecuniary	Sourcing	Revealing

Selling (outbound pecuniary innovation)

This covers gaining monetary value from unused technology or intellectual property, through selling or licensing. Interest was stimulated by a book entitled *Rembrandts in the Attic: Unlocking the Hidden Values of Patents* (Rivette and Kline, 2000), and the fact that some companies have generated substantial revenues: the Economist noted that IBM had developed patent licensing to a fine art, and had generated $1.7 billion dollars from its patent portfolio in 2001 (Economist, 2002).

Setting a value on unused intellectual property is not simple and one organization concerned with this is the Licensing Executives Society International, founded in 1965.

Sourcing (non-pecuniary inbound innovation)

Gaining ideas from publicly available information is daily business for scientists and engineers active in the early stages of innovation, for example through the scientific literature or from attending conferences. Monitoring competitor activities on the basis of their patents and announcements is also routine practice, and increasingly sophisticated tools allow analysis of patent portfolios (Stadlbauer and Drexler, 2014).

Acquiring (pecuniary inbound innovation)

The purchase or licensing of technologies by companies has become increasingly important as an alternative to developing their own technology. In an early article Chesbrough described this situation in terms of strategies adopted by Lucent Technologies and Cisco, both telecommunications companies competing in the same markets. Lucent Technologies incorporated much of Bell Laboratories (see Chapter 1; Gertner, 2012) and thus had very strong internal R and D expertise as a basis for technological innovation. In contrast, Cisco focused on external partnerships, acquisitions and investments to access technology, and became a powerful competitor (Chesbrough, 2003b).

A report is available which aims to support companies in buying in new technologies from external sources (Mortara and Ford, 2012). The main sections of this report deal with acquisition context, acquisition evaluation and acquisition options.

3.5 BUILD OR BUY?

Organizations face the choice between developing a technology in-house, and purchasing a technology from a vendor and this is referred to as a choice between "Build or Buy." There are many technical, business and strategic factors involved in such decisions which makes generalization difficult. One point to bear in mind is that while purchasing an external technology may sound attractive, it is difficult to judge the advantages and disadvantages without some in-house expert knowledge.

With respect to early stage innovation, one example of the choice between "Build or Buy" concerns technologies for managing ideas. Many organizations aim to stimulate ideas for improvements to products, processes, services or the organization itself. The classic "Suggestions Box," a physical box in which employees could deposit ideas on paper slips for improvements, has long been replaced by electronic methods, ranging from emails to sophisticated idea management systems. An organization thus needs to decide whether to develop their own idea management system, or to purchase one. Arguments in favor of an in-house system include detailed tailoring of the system to the requirements, and a desire to avoid being dependent on a vendor for upgrades and extensions. On the other hand, IT departments in an organization will not have the experience a vendor has accumulated from developing idea management systems for a variety of customers.

3.6 IS OPEN INNOVATION NEW?

The amount of attention which the idea of open innovation has received has inevitably led some to question whether or not the concept was new, with one author writing that "Most probably, open innovation practices are from all times" (Huizingh, 2010). In reviewing progress in open innovation since 2003, Chesbrough has defended the concept against criticisms that it is not new, or that it is new, but can be explained differently (Chesbrough and Bogers, 2017a).

It is, however, worth noting some examples of open innovation before the era of closed innovation. In a discussion of collective invention (Allen, 1983), it was argued that "many new production techniques have been developed by a process called 'collective invention.' When firms collectively invent, they make available to their competitors the results of new plant designs so that their competitors can incorporate extensions of those designs into new facilities they build." This was illustrated by an analysis of iron production in Cleveland, UK in the 19th century. Steady improvements in the design of blast furnaces were routinely shared with competitors. Allen suggests various reasons as to why companies might do this: (i) owners or managers of the companies might have had professional ambitions which were realized by releasing the information, (ii) to be seen as leading in the size and output of furnaces, (iii) it might have been very difficult to keep the information secret, given that the design and construction would have been carried out by contractors, (iv) there were ways in which it might have been profitable to release the information. Allen proposed that this mechanism of invention (which in the case of successful application should be termed innovation) is more general, and should be added to three other well-established mechanisms: via non-profit institutions such as universities or government agencies; through R and D by companies, and by individual inventors.

A much earlier example which involved this type of open innovation is provided by the Murano glass industry, originally established in Venice, but moved to the island of Murano, close to Venice, in 1291. The reason for the move, mandated by the city government, was the fire risks associated with predominantly wooden buildings close the furnaces required for glass production. However, the concentration of different companies in a small area, together with family connections between competitors, led to the rapid spread of new ideas (Johnson, 2014).

A third example is provided by the approach to research and development by the energy company Shell (Royal Dutch Shell), which was formed in 1907 by the merger of the Royal Dutch Petroleum Company of the Netherlands and the "Shell" Transport and Trading Company of the United Kingdom. It was clear that technical expertise was important, but the two companies took different approaches. Whereas the Royal Dutch Petroleum Company had established three laboratories in the Netherlands before the First World War, Shell Transport and Trading relied on external collaborations. To quote from the three-volume history of Shell (Jonker and van Zanden, 2007); "The importance attached to in-house research was a pertinent aspect of Royal Dutch policy almost from the start ... by contrast, Shell Transport relied on outsourcing any research the company wanted to get done." One example of the latter approach was a project commissioned in the chemistry department at Cambridge University, which demonstrated that a problem with kerosene from Borneo was due to the composition of the crude, which contained a relatively high proportion of aromatic hydrocarbons. This work was published in a leading journal. This example, in 1907, thus shows: (i) a strategic approach to open innovation, (ii) results which were important in practice, and (iii) a contribution to academic knowledge through publication.

While these examples are a useful reminder that open innovation can be found before the era of closed innovation, and that questions raised in Chapter 1 about innovation in general (how new? new to whom?) also apply to open innovation, there is no doubt that academic interest in open innovation as part of studies on how businesses develop and compete has been matched by a large variety of approaches developed in practice.

3.7 OPEN OR CLOSED?

Open innovation has attracted much attention from both academics and practitioners since Chesbrough's first publications in 2003. The reasons for this have been discussed (Huizingh, 2010): assigning a single term to a number of developments; timing, which coincided with interest in outsourcing, networks, core competencies and the internet; the many opportunities to develop the original ideas; and the integration of activities focused on accessing external knowledge with activities on external commercialization under the umbrella of open innovation.

The enthusiasm for open innovation has led some authors to conclude that the era of closed innovation was gone for good, as illustrated by the heading "Why closed innovation is doomed" in a short book on open innovation (Gaule, 2006). This is overstating the case. There may be good reasons for continuing with a closed innovation model, for example when there is a leading and protected position in developing an innovation, and the skills are available internally to achieve commercialization.

It is therefore better to consider, for any innovation-related activity, if progress would be faster, cheaper or lead to a better offering, by engaging with the outside world, and taking into account the costs and effort involved. Comparing the arguments for open or closed approaches across a range of projects will develop the capability to make assessments and help the decision-making process.

3.8 OPEN INNOVATION IN RELATION TO EARLY STAGE INNOVATION

Open innovation offers many opportunities to improve the early stages of innovation, particularly in the area of bringing in new ideas which can initiate or enhance potential innovations. Some of these are long established and well-known, for example attendance of scientific or technical conferences.

Much analysis of open innovation has been made at a company level, but it has recently been argued that more work on analysis of open innovation at an individual project level would be valuable. Some companies have extensively documented their open innovation activities, for example Philips has data on the open innovation activities of several thousand R and D projects (Vanhaverbeke, Du, Leten and Aalders, 2017b). It remains to be seen whether companies are prepared to collaborate with academics in analyzing this information, and what conclusions might emerge, and to what extent these will apply to other organizations.

3.9 OPEN INNOVATION IN PRACTICE

The following examples indicate the breadth of open innovation, with a particular emphasis on activities which support early stage innovation. Both the examples, as well as academic work on exploring and understanding open innovation, offer many opportunities for newcomers to open innovation to gain ideas on what might work well for their own organization.

Companies

Many companies have established mechanisms for engaging with the outside world in order to bring ideas in (inbound open innovation), recognizing the truth of what is sometimes referred to as Joy's Law. This states "no matter who you are, most of the smartest people work for someone else," and is attributed to Bill Joy, co-founder of Sun Microsystems.

Many of the examples of open innovation by companies are from multinationals, but there is increasing interest in understanding open innovation in SMEs, small and medium-sized enterprises (Brunswicker and van der Vrande, 2017; Vanhaverbeke, 2017c). Such companies, which typically have less than 250 employees, provide many of the jobs in the private sector, for example more than 60% in Europe. SMEs cover a wide range of companies including start-ups, spin-offs and long established family businesses. Brunswicker and van der Vrande describe four areas which merit further research in order to better understand open innovation in SMEs: crowd sourcing, network relationships, intellectual property, and management of open innovation.

The following examples illustrate how companies approach inbound open innovation. Many other examples could be found. It should be noted that these approaches evolve with time, according to experience gained or changes in strategy, and so the following should be seen as a snapshot at the time of writing (2017) rather than as definitive practices. An earlier description was provided in 2006 of open innovation at Proctor and Gamble, Qinetiq, Shell, DSM, Tate and Lyle (Gaule, 2006).

Major companies

Proctor and Gamble: Connect and Develop

Proctor and Gamble's program "Connect and Develop" is well-known and frequently cited as an example of open innovation. The change within the company from a closed innovation model to an open one has been described (Huston and Sakkab, 2006). A new strategy for innovation was developed from 2000, with the aim of generating half of the new products from P and G's own laboratories, and half from the P and G laboratories collaborating with the outside world. An early example of success is represented by the launch in 2004 of a new line of Pringle potato crisps, with pictures and words printed on each crisp. The idea for this had been generated in a brainstorming session in 2002, and one researcher conducted an early experiment using potato dough and her inkjet printer. Developing the concept required an ability to print sharp images, in a variety of colors, on crisps as they emerged from frying, at high temperatures and under high humidity. Solving these problems through an open innovation approach involved writing a technology brief which defined the problems, and circulating this through P and G's networks. This led to the identification of an Italian professor who also ran a small bakery, who manufactured baking equipment, and who had invented an ink-jet method for printing edible images on cakes and cookies. This expertise was quickly adapted to printing on potato crisps, and led to a successful new business. The use of an open innovation approach enabled this to be achieved more quickly, and at substantially less cost, than through an internal development program. The current version of the Connect and Develop program lists business needs for which P and G solicits solutions, from individuals, small businesses or major companies.

Nokia open innovation challenge

The 2017 Nokia open innovation challenge is the fifth in an annual series, and focuses on the Internet of Things (IoT) for smart cities, mobility, safety and security, and connected industries. Start-ups and innovators are encouraged to submit working prototypes which could lead to improvements in the above areas. The best entries will be developed in a two-day workshop in Helsinki with Nokia experts, followed by final judging and an awards ceremony.

This mechanism allows Nokia to access new ideas and the submitters the chance to co-develop their ideas with a major company.

IBM

IBM provides jam sessions, massive online brainstorming events, for both internal purposes as well as clients. The 2006 Innovation Jam brought together more than 150,000 people from 104 countries and 67 companies. From this ten new IBM businesses were launched with seed investment totaling $100 million (IBM, 2017).

In addition, IBM operates Centers for Advanced Studies (CAS) in 20 countries. These centers promote collaborative research, as well as educational projects, between IBM and local academic institutions. Encouraging innovation is one of the goals. One of the centers, CAS Benelux (Belgium, the Netherlands, and Luxembourg) focuses on using technology to innovate in the areas of earth, health and work environment, with a goal of becoming the "biggest driving force of innovation and research in the Benelux."

Shell: Shell GameChanger, Shell Technology Ventures, and Shell Technology Works

Shell has had collaborations with universities going back at least as far as 1907, as described above. More recently established mechanisms to facilitate open innovation include Shell GameChanger, Shell Technology Works, and Shell Technology Ventures (Shell, Innovating Together, 2017).

The Shell GameChanger process was started in the 1996 as a new way of driving innovation in the company, and has been described in various places (Hamel, 1999; Gaule, 2006; Ramirez, Roodhart and Manders, 2011). The program has evolved: the current program scouts for start-ups and businesses with unproven early stage ideas which have the potential to impact the future of energy. One recent example is the collaboration with a company called Kite Power Solutions, which generates wind power using two asymmetric kites. The kites fly in a looping, circular path and spool out over 100–200 m from a drum connected to a generator. While one kite is generating power, the second kite is being retracted. The advantages of this system include lower costs than other forms of wind power and optimization by flying at the height with the strongest wind. Progress in developing this technology has now attracted funding from Shell Technology Ventures, the corporate venture fund, as well as from Eon and Schlumberger.

A more recent open innovation program through Shell is Shell Technology Works, founded in 2013 and located in Cambridge, Massachusetts. The focus is on working with entrepreneurs and technology start-ups from outside the energy industry to meet needs in ways which are faster and more cost-effective than available alternatives.

Innovation through combinations of several major companies

In Chapter 2 the importance of diversity of inputs for generating new ideas was emphasized. This principle is illustrated in an example which also demonstrates early stage open innovation. Cisco, an IT and networking company with much experience in open innovation, hosts events known as Cisco Hyperinnovation Living Labs, or CHILL for short (Furr, O'Keefe and Dyer, 2016). Hosting the event involves detailed design of the event, and leading it. A key feature is that a simple short agreement has been developed to deal with intellectual property issues (discussed further below).

In one event the companies Airbus, DHL, Caterpillar and Cisco generated solutions to challenges derived from the combination of supply chains and digitization. Cisco has estimated that the resulting projects, spin-outs and joint ventures will generate $6 billion in revenues and save $3.4 billion in costs in the next ten years. Other Living Labs have dealt with various aspects of health, such as care coordination across different entities, virtual care, and medication management.

Small and medium-sized enterprises (SMEs)

The above examples focus on application of open innovation to early-stage innovation by major companies with a global presence and with long histories of innovation. Examples of open innovation by smaller, and newer, companies are less well documented, but are starting to emerge. One example is Ocean Optics (Brunswicker and van der Vrande, 2017).

This US company exploits opportunities enabled by new optical measurement technologies. The wide-ranging list of applications includes Farm to Table Technologies; Life Sciences; Environmental Monitoring; Authentication & Anti-Counterfeiting; Light/Laser/LED; Energy Technologies; Semiconductor Processing; and Materials Identification.

The success of an open innovation program of grants ("Blue Ocean Grants") to develop breakthrough innovations led to a second, larger, initiative called Open Photonics, which helps clients develop and commercialize new products, by providing services in product development, technology scouting, market intelligence and entrepreneurial mentoring. An open innovation model has been designed to connect clients with experts.

A set of nine case studies on open innovation in SMEs has been provided (Vanhaverbeke, 2017c). The companies involved are Devan Chemicals (textile technologies), Curana (bicycle equipment and accessories), DNA Interactif Fashion (application of technology to fashion, for example through 3D body scans), Dingens Barometers and Clocks (development of mercury-free barometers), Jaga (designer radiators), PRoF (an international open innovation consortium to develop new concepts in health care), and QOD (Quilts of Denmark, who adapted technology developed for NASA to create new bedding materials such as pillows and duvets). A further two case studies deal with collaborations between SME's and large companies: Isobionics, who produce flavors and fragrances using fermentation technology initially developed by DSM), and the Philips Airfryer, which allows food to be fried with much less fat, and which was developed by a small company called APDS.

These examples all have a technology component and all involve open innovation.

Competitions

A variety of competitions exist which have the purpose of stimulating ideas and innovations. Some are run by companies, with the aim of bringing in new ideas. Others are run by non-profit organizations, or supported by government agencies. Three examples of major competitions are given below, and many other competitions at a smaller level can be found, for example around specific technical themes.

Awareness of the scope and content of competitions may stimulate new ideas, suggest valuable contacts, or suggest ways of making progress in open innovation through organizing competitions. Establishing a competition should not, however, be undertaken lightly. There must be adequate resources available to screen, select, follow up on, and reply to the submissions. This becomes more difficult if the scale of the response is much higher than anticipated.

The Longitude prize

The current prize is a £10 million fund to be awarded for a diagnostic test that helps solve the problem of global antibiotic resistance, and is partly funded by a UK government agency. The competition opened in 2014 and runs for five years, with the prize money earmarked to develop and market the solution. The prize takes its name from the Longitude Act of 1714, through which the British Government offered a prize of £20,000 for a solution to the problem of establishing longitude in navigation at sea. This was a long-standing problem and essential to establishing safe routes to trading posts and settlements, an area of competition at that time between England, France, Spain and the Netherlands.

The Longitude prize illustrates competitions organized to meet societal needs, and a role of government in stimulating innovation.

The X-prizes

The X-prize foundation aims to stimulate innovative solutions to world problems. The first prize, of $10 million, was offered in 1996 for a privately-financed team which could fly a three-person vehicle into space twice within two weeks. As of 2017, there are eight prizes being actively competed for. The challenges are both technical, for example a lunar landing, or ocean bed exploration, as well as social, for example, adult literacy or women's safety. The number of teams competing for a prize in the present round vary between 5 (lunar landing) and 146 (artificial intelligence), and the prize varies between $30 million (lunar landing) and $1 million (technology for women's safety).

The innovations and partnerships developed by many of the teams may well contain useful information for those engaged in early stage innovation. In addition, one prize focuses on use of artificial intelligence to stimulate major innovations. To quote material on the X-prize web site (www.xprize.org), "The IBM Watson AI (Artificial Intelligence) XPRIZE is a $5 million competition, challenging teams globally to develop and demonstrate how humans can collaborate with powerful AI technologies to tackle the world's grand challenges. The prize aims to accelerate adoption of AI technologies and spark creative, innovative, and audacious demonstrations of the technology that are truly scalable and solve societal grand challenges. To encourage innovation in any form, the competition is an open challenge in AI. Rather than set a single, universal goal for all teams, this competition will invite teams to each create their own goal and solution to a grand challenge. The IBM Watson AI XPRIZE is a four-year competition with annual milestone competitions in 2017 and 2018. The top three finalists will compete for the Grand Prize at TED 2020."

Progress made by the 146 teams competing for this prize may well help other organizations with their efforts in collaborative innovation.

Katerva

Katerva is a non-profit organization. The name Katerva is derived from the Latin for crowd (caterva), emphasizing collaborative solutions. The focus is on global environmental and social problems which have the potential to be solved within ten years. Prizes are in the form of recognition as award winners, and are followed up by connecting the winners with mentors and investors.

Categories for individual prizes are, at the time of writing (2017): Behavioral Change; Economy; Ecosystem Conservation; Energy & Power; Food Security; Gender Equality; Human Development; Materials & Resources; Transportation; Urban Design; and Smart Cities. In addition there is an overall prize. Recent overall prizes have been awarded to SHE (2017); Ethereum (2016); The Ocean Cleanup (2015), the Khan academy (2014) and MBA polymers (2013). The Katerva process has three main steps:

1. A global network of 2500 spotters who identify potential major innovations
2. A proprietary analytical platform which selects the final 100 entries
3. Panels who evaluate the entries in terms of validation, scalability, policy and impact.

The Katerva process offers opportunities to identify, and engage with, promising ideas in a wide range of areas.

User feedback

User feedback is a relatively long-established mechanism for improving or generating new products, as illustrated by the 1978 article "Successful industrial products from customer ideas" (von Hippel, 1978). The beta-testing of products, particularly software, is also a way for a company to make improvements to their products before commercial release. Thus user feedback on a prototype is a form of early stage open innovation. In discussion of divergent views on open innovation, including the porous funnel view, the user-centric view, or the view that open innovation refers to the situation where all information related to the innovation is a public good, it was noted that an overarching, emerging theme is that of distributed sources of knowledge for innovation (Chesbrough and Bogers, 2017a).

Facilitators of open innovation

Accompanying the rise of interest in open innovation, there has been an increase in offerings which facilitate open innovation. The company GE offers GeniusLink, a way to connect problem owners with a very large number (over 21 million) of experts. Focus areas include engineering, design, IT and marketing. IBM provides IT infrastructure to support open innovation and Cisco, as noted above, hosts CHILL events involving multiple organizations.

Companies which match problem owners with external providers of solutions are sometimes called knowledge brokers. One well-established example is Innocentive, formed in 2001 within the pharmaceutical company E.I. Lilly, and spun out in 2005. The company operates by connecting problem owners (called "seekers") with a network of experts who could provide solutions (known as "solvers") and providing support to the formulation of the problem, IP issues and financial rewards for the solvers. Many scientific and technical disciplines are covered, as well as business, entrepreneurship and social innovation. Innocentive is also partnering with others: for example, a partnership between Innocentive and the Royal Society of Chemistry helps application of RSC members' knowledge.

Governments facilitate open innovation, for example by providing funds for consortia. Such consortia can focus on pre-competitive research (a form of early stage innovation) between companies, or between companies, universities and other research institutes such as government laboratories. These non-commercial research organizations are sometimes referred to as knowledge institutes. A major, and early, example is that of Sematech (Semiconductor Manufacturing Technology) which was founded in the 1980's to solve common manufacturing problems across different US companies, and to enhance the competitive position of the US semiconductor sector in the light of increased international competition. The consortium received $100 million per year of US government funding (Mazzucato, 2014). Many other consortia have been established, for example the Dutch Polymer Institute, established in 1997, lists, in the 2014 annual report, 46 knowledge institutes and 37 commercial partners. For consortia which depend on research by PhD students it will often be desirable to have at least a second round of funding, for a new set of PhD students, in order to maximize benefits from the research programs and networks which have been established.

A report entitled "An innovation-led boost for US manufacturing" (Sirkin, Rose, and Choraria, 2017) notes that in addition to the example of Sematech, fourteen consortia have been established since 2014 by NIST (National Institute of Standards and Technology), a US government agency whose mission is to promote US innovation and industrial competitiveness. The report also notes that the record of US consortia has been mixed, and makes recommendations for improvements.

Universities

Universities participate in open innovation in a variety of ways, through collaborations with companies in the form of individual projects, or through consortia, as noted above, and through technology transfer offices. Related activities, which may be more or less open, include science parks or incubators for start-ups derived from academic research.

Early stage technical innovation frequently involves interactions between research departments in universities and companies. New recruits to a company, with a recent research degree, bring valuable relationships, contacts and up-to-date scientific knowledge to a corporate environment and are often involved in establishing projects with universities. In addition, senior technical staff in company R and D departments may hold part-time appointments in universities, or be active in large collaborative efforts such as multi-party consortia, involving a range of universities, other research institutes and companies.

Companies will be primarily interested in obtaining new scientific results and insights from collaborations with universities, on topics which do not justify a purely internal effort. In some cases the collaboration exists to explore a new technical area; in other cases the collaboration may provide background to an internal program. In addition to obtaining results, companies will also be interested in the following aspects of collaboration:

Speed in setting up and executing collaborations

Large corporations can be slow due to extensive internal bureaucracy. This does not necessarily mean inefficiency. An R and D proposal may well have to be reviewed from technical, business, contractual and IP perspectives, each of which involves separate (and busy) departments. However, universities can also be slow to respond. While it may be possible for individual professors to establish contracts, in other cases a technology transfer office may act as an intermediary, and in addition there may be internal review boards whose approval is required, for example at departmental, faculty or university levels. The collaboration can also be slowed by time needed to identify and put in place a suitable graduate student, and, after the research is completed, by the time needed to complete the writing of a thesis. It is not unknown for these two steps to each add a year to the actual duration of the research. Thus for a PhD program of 3 years, the total time involved might be 5 years, and corporate interests may well have changed in this period. In addition, corporate staff may have moved to other positions during this time, which gives a lack of continuity.

Access to networks

The primary contact in a collaborative research project between a company and a university is through the research student and his or her professor. Additional contacts may well be valuable, for example with colleagues in the same department, or (inter)national colleagues in the same technical area. These contacts may provide stimulating discussions which increase the benefits of the collaboration and may lead to additional collaborations.

Recruitment

A collaboration of several years duration provides a good opportunity to assess a research student as a potential employee, taking into account not just the qualities as a researcher, but drive, initiative and communication skills. Some students will of course prefer to pursue other career paths, for example within academic research. The collaboration also helps other students not directly involved in the collaboration to be aware of what the company might offer in terms of opportunities.

From the university point of view, the primary attraction of collaboration with industry is additional funding for research. Other factors also need to be taken into account:

High quality research theses

Academic research is concerned with the production of theses and publications. The interests of a sponsoring company may range from obtaining new scientific insights to generating data which is directly useful to the company. In the latter case it must be recognized that a PhD is an original contribution to knowledge, and not just the generation of data which does not add to the value of the PhD. In other words, industrial sponsors must be careful to bear in mind whether requests for experimental work are consistent with the production of a high quality PhD. If not, it may be possible to establish a different contract, focused on producing a particular set of data, perhaps carried out by technicians or as an undergraduate project.

Perspectives on the industrial context

A collaboration between a university and a company offers the former the opportunity to understand the context of the collaboration from an industrial point of view. Awareness of the time-scales, commercial incentives, different skills, and effort required to bring new scientific research into commercial reality may well be stimulating, and provide valuable teaching material at a non-confidential level. Different technical areas, for example science and engineering, can have little overlap in an academic environment, but will be routinely represented in the same meetings within a commercial R and D organization. Exposure of academic staff, and students, to the interplay of different backgrounds can be stimulating.

Conferences

Scientific and technical conferences in which the latest results are presented are very well established and routinely attended by academics, students and industrial researchers, all of whom may be active contributors. This awareness of new developments can influence the direction of industrial research, lead to new internal programs or to new collaborations between universities and companies. This is thus a good way to facilitate early stage open innovation.

Further opportunities are offered through a role in conference organization. An industrial participant in the committee responsible for selecting a program, and inviting contributors, at a scientific conference can help to highlight particular themes, and develop background information. Inviting key individuals from academia, industry or government to present keynote lectures (many will accept) offers the opportunity to create and build new relationships. Publishing conference proceedings enables useful information to reach the public domain.

3.10 CHALLENGES IN OPEN INNOVATION

The above has made clear that open innovation has become an important part of innovation, and new examples in practice continue to emerge, as well as new insights in understanding the subject. In the context of this book the main impact of open innovation is on companies accessing ideas and skills in the outside world, in other words in-bound open innovation. There is little doubt that open innovation is here to stay, although it has been suggested that the term might disappear as open innovation becomes standard practice (Busarovs, 2013).

It should, however, be recognized that open innovation does involve some significant challenges, which are not always appreciated by the more enthusiastic proponents of the subject. The time and money spent on addressing these challenges has to be weighed against the benefits obtained. The following three aspects will frequently have to be addressed, preferably at the start of an open innovation project and then modified in the light of experience. Starting open innovation without some assessment of the following will inevitably lead to problems.

Effort involved

Establishing collaborations, either on a 1:1 basis or in the form of multi-party consortia, takes considerable time. Companies will require business and legal approval for the content of the project as well as for the contracts. Researchers in universities may have to work through technology transfer offices or require higher level agreements. Junior staff in a company, assigned to establish external projects, may find it challenging without support in terms of training or mentoring.

Setting up competitions, or inviting the general public to contribute solutions to a problem of business interest to a company, or even running large scale internal events may require significant resources to process the results. The IBM Innovation Jam session described above brought together 150,000 people from 104 countries and 67 companies and resulted in the launch of 10 new IBM businesses with initial funding of $100 million. Organizing and funding an effort of this magnitude is clearly outside the scope of most organizations, but the example does illustrate the point that effort required to follow up on the results of an event of this type must be considered at the outset.

Intellectual property

Intellectual property includes trademarks, copyright, design rights, and patents, but out of all these, patents are most likely to concern those involved in early stage innovation. Given the legal nature of the subject, the following is a brief introduction, and more definitive information should be obtained in practice from patent lawyers in the appropriate organization, taking into account the patent system of interest. The criteria for obtaining a patent vary somewhat according to which legal system is used, for example a national, regional (e.g. Europe) or world patent application, but in general include the following:

— the subject should be patentable. While this is not in doubt for many technical areas, inventions in areas such as software, business methods or (modified) life-forms require closer examination.

— novelty. This means that the idea or invention has not previously been described. This can be in any publication, not just reputable scientific journals, and this body of existing knowledge is referred to as the "prior art." There are many scientific results which do not justify the further work necessary to develop into a publication or patent, but, by being disclosed, become part of the prior art, and thus prevent patents being subsequently based on the finding. A commercial organization, Research Disclosure, publishes, for a fee, research findings in a paper journal and an online database.

— non-obvious or inventive. The idea here is that the result should not be obvious to an expert in the field (referred to as "one skilled in the art"). It can be difficult for those with recent PhDs, which are based on clear thinking and logical explanations, to be confronted with this requirement. Fortunately patent lawyers can be quite skillful in turning scientific results into patent claims, (as well as being able to dismiss scientifically interesting results as un-patentable).

— useful. This criterion requires the subject of the patent is capable of being applied, and in a beneficial manner. This is not usually an issue, but may arise when an invention requires a formal approval before it can be applied, for example in the case of new drugs.

Discussions on how patent rights should be assigned are particularly troublesome in the context of collaborations between partners, and typically is a point of discussion between companies who develop an idea which was derived, wholly or in part, from an external collaboration. At least three erroneous assumptions lie behind some of these discussions:

— firstly that patents automatically represent value. This ignores the fact that many patents remain unused. Some patents may be applied for in order to give a company more time to formulate a business and technology strategy; others may be applied for in order to hinder a competitor.

— secondly, that patents give the right to practice the findings covered by the patent. This is incorrect. A patent prevents someone else from using the findings covered by the patent. Whether or not the owner of a patent can use the invention may well depend on whether other patents block it. For example, a patent on a particular component, or step, in a manufacturing process might be invalidated by a patent on the process as a whole.

— thirdly, that the revenue generated by the innovation (defined in Chapter 1 as the successful exploitation of new ideas) is due in a significant part to the original idea. While some patents generate significant royalties, many do not. Major new processes may involve many patents. As described above, the Shell Gas-to-Liquids process is a proprietary process covered by 3500 patents. Assigning value to one of the many patents, taking into account the development costs (over $18-19 billion for GtL) would be a formidable task, if not impossible.

In addition, it is not always recognized that patenting is expensive. Annual fees have to be paid, which clearly increase with the geographical coverage required.

It is sometimes suggested that assigning patents to both partners in a collaborative venture is a solution. While this might work in some cases, it is likely to require considerable work to establish how the benefits accrue to each party, and the freedom of each party to license the patent to others.

Finally, it should be noted that there is a debate about whether patents stimulate or hinder innovation. It has been argued that the patent system does not favor innovation: firstly, there is evidence that increasing the scope of permissible patents in US agriculture in the 1970s did not lead to improved agricultural productivity; secondly, the patent system leads to companies or individuals acquiring patents for the sole purpose of selling them at a higher price, generating a "parasitic ecology of trolls and defensive patent holders"; and thirdly, the patent system is expensive and creates substantial costs, for example for patented drugs rather than generic ones (Economist, 2015). Whatever the merits of these arguments, it is unlikely that the patent systems around the world will change quickly.

Managing relationships

While some collaboration can be managed on a straightforward basis of money in exchange for data, others will require the establishment of trust and good levels of communication to maximize the benefits (see Chapter 4). There may be grey areas in terms of what information should be shared: academics will not want to pass up the opportunity of obtaining novel scientific results in sponsored research even though the direction might be tangential to the agreed main program. Companies sponsoring research will not want to reveal the full strategy and commercial potential of the research. In the case of large companies collaborating with small companies the former will want to make the potential value of the collaboration attractive enough to convince the small company to commit to the collaboration, but not so large that claims on future revenue streams appear unreasonable. These and other issues can emerge in the course of the collaboration but it is easier to deal with them by having some discussion at the start of the collaboration on expectations from both sides, including on criteria which make the collaboration successful, and possible pitfalls.

In the case of collaborations between large and small companies, it should be recognized that the former have much larger resources in, for example, legal and IP departments, as well as wide-ranging expertise in collaborations built up over many years. In contrast, small companies may rely on an R and D director, with a fraction of this experience, to agree to contracts and IP rights. Large companies should endeavor to treat a smaller partner fairly; the latter may want to bring in expert advice.

A large organization which engages more with the outside world for ideas or on research projects may have to manage internal relationships. Accessing skills outside the organization which are more extensive, and are cheaper than those available inside the organization can be perceived as a threat to the professional status, or even usefulness, of internal experts. To avoid creating unnecessary friction it is wise to involve internal experts in efforts in open innovation. This can range from initial advice on selecting partners up to managing the project.

3.11 RECENT LITERATURE AND TRENDS

The above demonstrates that open innovation is a very active area, both in terms of practice and in terms of academic research. It seems clear that open innovation, far from being a temporary fad, is here to stay.

Four recent publications provide a much more extensive treatment of the subject of open innovation and the following outlines give areas of current interest:

1. A multi-author volume entitled, *New Frontiers in Open Innovation* (Chesbrough, H., Vanhaverbeke, W., and West, J., 2017c) summarizes progress to date in understanding open innovation, and discusses in fourteen chapters topics which merit further research and understanding in the coming years. These topics are grouped into four parts:
    - Open innovation: 10 years later.
    - Analyzing open innovation at different levels of analysis.
    - New application fields for open innovation.
    - Managing and organizing open innovation.
2. A second book, published for the Product Development and Management Association, is also a multi-author book, focusing on practical aspects and examples (Noble, Durmusoglu, and Griffin, 2014). The twelve chapters are grouped into five parts:
    - Open innovation in the fuzzy front end.
    - Open innovation in the development stage.
    - Open innovation with universities.
    - Open innovation for really big initiatives.
    - Best practices and advice for open innovation.
3. A third book (Vanhaverbeke, 2017c) specifically addresses managing open innovation in Small and Medium-sized Enterprises (SMEs). The author argues that open innovation in SMEs, although much less well documented than for large companies, is substantially different, for four reasons: firstly SMEs often do not have a portfolio of innovation projects, managed through a funnel (Chapter 2); secondly, open innovation in SMEs is much more directly linked to strategic changes; thirdly the management of open innovation is different in SMEs compared to large companies (for example by being directly the responsibility of the founder) and fourthly, management of networks tends to be different,

with much closer personal contacts, faster decision-making and more informal communications than with big companies.

Case studies (noted above) are included, and the nine chapters address:

- Open innovation in SMEs
- Business model innovation in SMEs
- A dynamic view of business model innovation
- How SMEs create and capture value through open innovation
- Managing innovation networks
- How can start-ups collaborate with large companies?
- Two examples of how SMEs collaborate with large companies
- Working effectively with innomediaries
- How to apply open innovation in your company

4. A short article entitled "The future of open innovation" (Chesbrough, 2017b) notes three areas which will be important in the future of open innovation. Firstly, business model innovation, designed to utilize knowledge flows into or out of an organization, and linked to technology innovation. Secondly, open innovation in relation to services, which are becoming a major source of revenue for companies. Services include leasing products rather than selling them; GE does both with aircraft engines. Thirdly, open innovation at the level of a community of participants, as exemplified by Taiwan Semiconductor Manufacturing Corporation, which manufactures chips designed by clients, and which is managed through an open innovation platform.

The themes covered by these four publications give recent practitioner and academic views on open innovation, provide an overview of the progress that has been made, and indicate directions which are likely to be explored in the coming years. Combining this information with the points discussed in this chapter gives a good basis for individuals or organizations embarking on open innovation, as well as providing examples and insights which may be helpful to those already involved in open innovation.

4 Behaviors, Teams, Communications and Learning

4.1 INTRODUCTION

Two major components of an environment which lead to innovations are firstly having the right processes to identify and develop ideas — as discussed in Chapter 2 — and secondly having the right organizational culture, in particular with respect to behaviors, teams, communications and learning.

Organizational culture is a topic which has been much studied, and has been defined as "the types of attitudes and agreed ways of working shared by the employees of a company or organization" (Cambridge Dictionary, 2017), and "the values and behaviors that contribute to the unique social and psychological environment of an organization" (Business Dictionary, 2017). Here we focus on the culture in relation to early stage innovation, and view it simply as the sum of behaviors in those parts of the organization directly concerned with innovation. It can be readily appreciated that the differences between a culture which is hostile, indifferent or supportive of innovation will make a major difference to the success of innovation efforts, whatever the formal commitment or slogans are. It should be recognized that different parts and levels of a large organization often have their own cultures, from those in whole divisions down to individual departments or teams.

The importance of behaviors and communications to the subject of this book is supported by the following:

— in many of the case studies on early stage innovation presented in Chapter 5, comments are made on the importance of these factors to the success of the innovation.

— a survey in 2015 of a wide range of companies across different sectors, covering 200 respondents, led to the most critical innovation challenge being identified as "problems with organization/organizational culture or mindset." (Imaginatik, 2015)
— the quotes at the start of this book by Negroponte and Machiavelli which indicate that innovators can be difficult to deal with, and that innovation can breed resistance. Clearly the five hundred years since Machiavelli have not seen these issues disappearing.

In addition, various studies have emphasized the importance of generic skills such as teamwork and communication in pharmaceuticals, fine chemicals and biotechnology, with employers increasingly calling for training in these areas to be included in undergraduate courses (Peplow, 2017). It was noted in this article that while many undergraduates take part in group work, teamwork is different: the former typically involves classmates contributing to the same goal, whereas teams within a company are often made up of people with very different areas of expertise, which thus demands greater skills in communication.

Topics such as organizational culture, teams, behaviors, communications and organizational learning are not ones which typically have immediate appeal to newly-graduated scientists and engineers working in a technical environment. Nevertheless, a common experience is a growing realization, over the course of a career, of their importance. This chapter aims to provide a brief introduction to ideas and literature which will help those working in innovation within an organization.

4.2 TWO IMPORTANT BALANCES

Individuals, and teams, focused on early stage innovation are faced with two key balances to maintain. One is concerned with efficiency versus effectiveness, the other with enthusiasm versus objectivity. The question of whether these factors are optimally weighted can easily be left unaddressed in the effort to produce results, but regularly assessing the performance of an innovation team with respect to these factors can prevent problems developing.

Efficiency versus effectiveness

In a business environment there is, rightly, much emphasis on efficiency, and this is particularly true of operations such as those involved in running expensive equipment or manufacturing plants. Efficiency is concerned with maximizing outputs in relation to inputs such as money, time and use of equipment. Efficiency becomes of over-riding importance in crisis situations, for example when it is necessary to implement stringent measures to avoid imminent financial collapse or — more acutely still — when a building needs to be evacuated in the event of fire.

Effectiveness, on the other hand, is concerned with establishing the best idea or solution, without the pressure of constraints such as time being the dominant factor. That this can require rigorous thought is illustrated by a quote attributed to Einstein: "If I had an hour to solve a problem I'd spend 55 minutes thinking about the problem and 5 minutes thinking about solutions." In addition, it might be appropriate to gather a wide range of inputs before defining and executing an innovation project, which also takes time.

The differences between an approach based on efficiency and one based on effectiveness often become apparent in idea-generation workshops, or problem-solving sessions, which bring together participants from operational and research backgrounds. The former are more likely to jump into defining actions; the latter are more likely to question the definition of the issue being considered, and to pursue divergent options. Explicitly recognizing such different approaches, and the fact that both are necessary, helps avoid irritations arising.

Enthusiasm versus objectivity

Many new ideas require a champion with enthusiasm, and the power to persuade, in order to drive them forward, and obtain support and resources. Such champions often have strong and visible personalities, but this is by no means always the case: some projects make considerable progress by the proponents keeping a low profile, either from a conscious desire to avoid attracting too much attention to the project, or because of the styles of the team members. Projects may also have different levels of visibility, independent of their proponents' styles.

In all cases, however, enthusiasm for a new initiative has to be complemented by objectivity. This applies both to project champions and teams, who should not be blind to weak points in their case, as well as to their colleagues, managers and sponsors, who need to ensure that resources are not being wasted. For an individual champion, it can be helpful to find a colleague, or a mentor, with whom difficult aspects can be discussed. Within a team, someone can be asked to play the "devil's advocate" and ask the tough questions before others do (for an example see the contribution by Prof. Thompson in Chapter 5).

When attempting objective evaluation of enthusiastically championed proposals, it is not enough to focus purely on the arguments involved. Attention must also be paid to the style of questioning. Asking hard questions in an aggressive manner (sometimes with an underlying intention to demonstrate superior knowledge or intellectual ability) is unlikely to motivate the "victim" to revise the proposal, or to submit another one. An organization which effectively destroys all new suggestions will not succeed in any ambition to be innovative. On the other hand, resources should not be made available to support just any proposal without questioning (see also the discussion of stage gates in Chapter 2).

Getting the right balance between being enthusiastic and being objective is made even harder in an era of globalization, open innovation, and teams of different backgrounds. What might seem healthily robust conversation in one country might be interpreted as being unnecessarily blunt or rude in another, and such perceptions can be further influenced by differences in age, gender and experience.

As in the case of efficiency and effectiveness, it is worthwhile to periodically take time to reflect on experiences with enthusiasm and objectivity, and identify where improvements can be made.

4.3 TEAMS

Understanding and developing teams has been a significant topic in many contexts (e.g. in business, sport, or the military) for many years, and continues to be so, as reflected in recent articles with titles such as "The secrets of great teamwork" (Haas and Mortensen, 2016). The case study provided by GE (Chapter 5) describes the challenge, and rewards, of establishing multi-disciplinary global teams at a time (mid 1990's) when this way of working was not as common as it is today. At about that time, team work in an era of globalization was discussed, posing the question of whether teams that did not spend time physically together could be effective. The answer was yes, provided ways to build credibility and trust could be found (Bensen-Armer and Hsieh, 1997).

Teams dealing with innovation face two specific challenges: (i) the high degree of uncertainty about the outcome of early stage innovation; and (ii) the high degree of enthusiasm which proponents of an idea often have. The combination of these two factors can lead to misunderstandings or conflicts within innovation teams, or between innovation teams and their stakeholders.

It is important to build trust and credibility within the team from the outset. This requires some effort, and may also seem like a waste of time to those anxious to get on with projects. However, this initial investment of effort and time is well worthwhile, since it is easier than repairing an established team which has become dysfunctional. Two approaches which are often used to support team building are Belbin team roles, and personality tests. In both cases, the aim is to clarify individual working styles and preferences, leading to insights on how the members can best collaborate. This may also contribute to selection of new team members, although for teams made up of specialists there may be limited choice.

Belbin team roles

Belbin studied individuals in teams, and published the research in a book in 1981. The main conclusion was that an effective team has nine roles which need to be covered. The roles were named Coordinator, Shaper, Plant, Monitor Evaluator, Implementer, Team Worker, Resource Investigator, Completer Finisher, and Specialist. Further description of these roles, and in relation to innovation, have been discussed (Hender, 2003; Mostert, 2015). The monograph by Hender primarily focuses on innovation leadership (see below) but discusses roles in the context of leadership issues. Mostert investigated how Belbin Team Roles apply to innovation teams in an R and D environment, and carried out research based on a comparison of 730 responses from different teams, mainly in the UK and the Netherlands, whose members had completed the Belbin Team Role Self-Perception Inventory. The main conclusion from this work was that all individuals and teams had their own specific mix of Belbin roles, and hence each team had a unique innovation culture and climate. However, a common feature was that team members benefited from knowing each other's Belbin Team Role and the effect of their own roles on the teamwork.

Personality tests

Personality tests are widely used to improve collaboration and team work, as well in recruitment and career development. Of these, the Myers-Briggs Type Indicator is the most popular. Based on a self-completed questionnaire, one of 16 personality types is assigned, based on preferences for modes of perception, judgment, source of energy, and assessment of situations. The language used in MBTI differs somewhat from everyday use, for example extraversion refers to those who get their energy from the outside world, whereas introversion refers to those who get energy from their own thoughts and activities. In a short article on the history of personality tests (Harrell, 2017), it was noted that critics of MBTI argue that it does not predict individual or team effectiveness, and that more than half the people who retake the test get a different result the second time. Others, however, find that it helps understanding own styles and preferences, as well as those of colleagues, and that it helps to reduce conflict. The article by Harrell also briefly discusses other personality tests such as the Five-factor model, or Strengthsfinder, and noted an increasing interest in custom-made tests.

Feedback

Many organizations give regular appraisals of staff performance, either as part of a structured process or informally. Giving feedback is a routine component of such performance reviews, but it is also a way to help teams to develop. Formulating and expressing views about what a colleague could do more of, or less of, can be valuable, as well as sharing perceptions of each other's strengths and weaknesses. Exploring the weak points of projects within a team is a useful way to prepare for external criticism.

Conflict in teams

It may well be true that "for good ideas and true innovation you need human interaction, conflict, argument, debate" (a quote attributed to business leader, author and speaker Margaret Heffernan) but this requires, like diversity, an ability to manage the differences. It is therefore worthwhile to have some ideas on how to manage team conflicts, and on understanding different styles and personalities, before relationships in a team deteriorate. Teams composed of experienced members may know, through direct experience or through observation, ways of avoiding and resolving clashes, whereas others may have to learn the hard way.

One approach advocates holding a series of conversations when a new team forms, focusing on five topics: how people look, act, speak, think and feel (Toegel and Barsoux, 2016). A variety of questions for discussion are suggested, such as: "in your world, is a promise an aspiration or a guarantee?". These help explore differences without the pressure of a project not going well. This becomes particularly relevant when there are significant differences in professional or personal background. It may be valuable to have meetings, at least initially, facilitated by someone outside the team with experience of the issues which can arise, but who is not directly involved in the content.

Leadership of innovation teams

While junior staff who join a new organization are perhaps unlikely to start in a role leading an innovation team, it can be useful to reflect on the style of the leader, and some literature helps to put this in a broader context. A short monograph on innovation leadership (Hender, 2003) covered:

- models, structures and roles in innovation;
- types of innovation leader (described as intrapreneurs who turn ideas into realities inside the organization, entrepreneurs who develop the product or service into a money-making proposition, and extrapreneurs who manage the boundary between the new ventures and the core business);
- supporting roles (the innovation team, the sponsor, the specialist and the sounding board);

- imperatives (key tasks for innovation leaders), which dealt with creating a climate for innovation, recruiting for innovation, team building, managing innovation teams, rewarding innovation, developing skills, and finding and working with supporters;
- examples, including case studies of 3M, easyGroup, and British Airways.

In a study analyzing Belbin Team Roles in innovation teams in an R and D environment, it was found that many of the team leaders were "Shapers," described in Belbin terminology as challenging individuals who "provide the necessary drive to ensure that the team is kept moving and does not lose focus or momentum" (Mostert, 2015). While this can be stimulating for the team members, they sometimes experience the Shaper as being too pushy; in this case cooperation with someone with a Coordinator profile can be beneficial; a Coordinator focuses on the team objectives, draws out team members, delegates work appropriately and acts as an interface between the Shaper and the team.

An article on how managers empower their organizations to innovate emphasized the point that "innovation is a process of discovery that requires a different approach to leadership than the core business," and that success as an innovation leader comes more from example than by dictating, by asking questions rather than making decisions, by removing obstacles, and ensuring that people have the right time, constraints and tools (Furr and Dyer, 2014).

Teams and creativity

While new ideas, or initial thoughts, often originate with individuals, groups in the form of teams or workshops are frequently involved in developing ideas, or creating new ones, or finding ways around barriers. Creativity in teams is thus an aspect to be considered, and this was discussed in Chapter 2 with respect to idea generation.

With respect to behaviors one author has noted, in a discussion about how to kill creativity, that creative teams should encompass diversity, a shared excitement about the goal, willingness to help each other through difficult periods, and recognition of unique knowledge and perspectives within the team (Amabile, 1998). These points merit some further observations.

Diversity

Diversity is generally considered an advantage for creative teams, with its potential for stimulating new opportunities and solutions. It is, however, important to recognize two additional points. Firstly, ensuring diversity at the level of professional background, gender, age, ethnicity, country of residence, etc. does not necessarily mean that a diversity of thinking will follow. This is a particular danger when a strong corporate culture is in place, which leads to employees with similar attitudes. In setting up teams, it is worth checking whether diversity of thinking is present, rather than just a superficial diversity.

Secondly, if diversity of thinking has been established, there will be a need for skills in the team to constructively manage the differences. A range of opinions is unproductive if it simply leads to conflicting views which are left unresolved, or resolved in a way which is unsatisfactory to some team members.

Shared excitement about the goal

The goals of early stage innovation are usually motivating to those taking part. If, however, the objective does not seem worthwhile then it may require further definition, for example to give clarity on the reasons to achieve it, or to obtain clear management endorsement.

Willingness to help each other

While it might be expected that this would be a given, in practice individuals may have very different aims in what they want to get out of a joint project. Team effectiveness is improved by creating an atmosphere in which people feel comfortable sharing personal goals. It may well be that a member of a team feels that full disclosure of his/her ambitions is not in their own interest, for example if they relate to future job moves or promotions, and this of course depends on the degree of trust amongst team members. Nevertheless, linking the goals of the organization to goals of the team and then to goals of team members is useful, and will help team members to achieve personal goals as well as the team goals. One way to do this is to display, for example as posters, a cascade of goals from high level to individual.

Making use of skills

In forming a new team it is usual for people to introduce themselves on the basis of their current job, and perhaps initial training. It may well be however, that other, earlier experiences, contacts and skills which do not seem relevant at the outset become so if the project takes an unexpected turn. It is thus helpful to not only share backgrounds at the start but to review them regularly for relevance, and for possible productive combinations.

Challenge versus support

The need to balance challenge with support to help students learn was formulated by Nevitt Sanford in the 1960's, and has since been used to describe teams in other environments. Figure 4.1 shows the four quadrants which Sanford generated by placing degree of support for each other in a team on one axis, and degree of challenge to each other on the other axis. A team will perform well when in the top right quadrant: here there

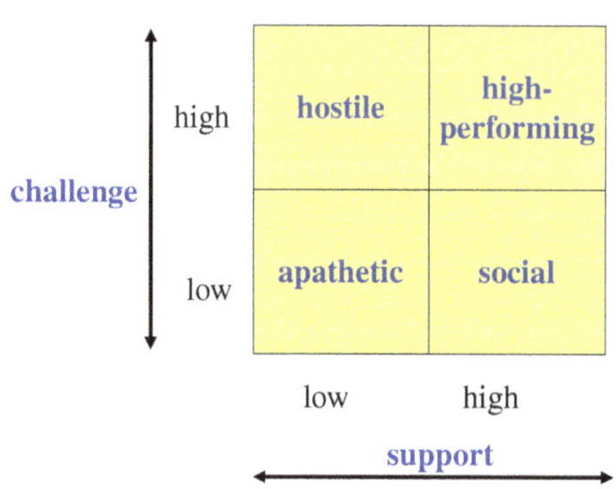

Fig. 4.1. Challenge and support in a team.

is sufficient trust and support within the team to make challenging questions possible, without seeming threatening. If the support is high, but the challenge is low, then the team will accept each other's contributions without much questioning of outcome or effort, and the team will have more of a social character. Challenge without any sense of trust or support is likely to create a hostile atmosphere, whereas if neither challenge nor support is present, the team will merit the designation of apathetic.

This matrix is useful in several ways. Firstly, in asking a team to assess in which quadrant they best fit, and establishing if perceptions are the same or different across team members; views from people outside the team can also be included. Secondly, if not already in the high performing quadrant, a team can consider ways of building in more challenge, or establishing more trust and support. Thirdly, the matrix allows people to reflect on past teams of which they have been members and to use both good and bad experiences as examples.

The need to balance challenge with support has been reinforced, using different language, in a study conducted by Google to establish the factors which led to effective teams in their organization (Rozovsky, 2015). More than 200 interviews were held over two years, and more than 250 attributes of 180 teams were examined. The initial assumption was that this study would identify the perfect mix of traits and skills necessary for a high quality team. Instead, the study showed that successful teams scored highly on the following points:

- psychological safety
- dependability
- structure and clarity
- meaning of the work
- impact of the work

Of these five factors, the first was found to be the most important by far. It covered being able to ask what the goal was, when it was assumed that everyone knew this, or asking for clarifications which could be seen as being uninformed. This study led to a new tool being created to help teams become more effective.

4.4 COMMUNICATION ISSUES

The primary experience of most people in communicating about their work during their scientific or technical training is usually with their peers. This covers presentations within their own department or organization, as well as oral or poster presentations at meetings and conferences. Written reports include those required by those funding the work, as well as publications in journals.

Communication within an organization requires a wider range of topics and stakeholders to be addressed. Examples of topics which will require attention are:

- strategic fit: where does the project fit relative to the overall goals?
- business implications: when, and by how much, will the project, if successful, contribute to new income streams or cost reductions?
- peer understanding and challenge: technical comments and questions from within a particular discipline, as well as related disciplines required for implementation. For example, a Proof-of-Concept which has been established with a basis in chemistry or catalysis will sooner or later require detailed discussion with chemical engineers on scale-up.

Getting the breadth and depth right across these topics requires careful thought, which sometimes is omitted under the pressure of time. It also requires taking into account the interests of the particular composition of the audience or readership.

Communicating about early stage innovation is made more difficult by the high level of uncertainties around the outcome, the costs and perhaps even the strategic fit. On the other hand, starting with a relatively small commitment of resources means that the degree of challenge and scrutiny will also be small. As the project progresses, the uncertainties decrease, but the costs increase, and the evaluations and challenges become more rigorous.

Recognizing, and being explicit about the uncertainties, will facilitate the effective communication of projects. In particular, as discussed in Chapter 2, formulating key project unknowns, whether technical or market-based in nature, and planning activities to address these unknowns will help to create clarity and generate understanding.

In practice, there are a number of points to consider when developing communications:

Channels

There are many different ways of communicating, including formal reports, newsletters, briefing notes, presentations and informal discussions. Attention should be paid to the communication preferences of the recipient: someone who values a solid, written report will be less impressed by an informal chat, and vice versa. Such preferences can be individual, or relate to the preferred style within an organization. It is thus worth reflecting on whether the means of communication, and the style, is the most appropriate.

Anecdotes and stories

Anecdotes and stories are a key feature of organizational life. Depending on the speaker, they can be detailed or sketchy, lengthy or brief, and include gossip, history, or entertainment. Some may amount to 'advertising' for the speaker or his/her unit. While these goals have some validity, the better anecdotes or stories are the ones which contain a lesson or message which can be applied by the listener, in what are probably different circumstances.

Organizations offer many examples of different styles in communication and this offers opportunities to reflect on what is effective and why, taking into account that most people will not want to adopt a radically different personality for the sake of communication. In addition to observation, training is widely available on improving communication skills, and may be required for more senior positions involving commercial contact with the outside world, or for leading change internally.

"Elevator pitches"

Elevator pitches are short verbal presentations. They are based on the idea of meeting someone influential, for example a senior executive, in an elevator, and having the duration of the elevator ride to make good use of this opportunity. In practice this situation will rarely arise, but thinking through what to say in such a situation, and how to say it, is a useful way of providing structure which is effective in a wider range of circumstances. Pitches are often associated with a request for funding, but advice or contacts can also be sought. However, the main point of constructing an elevator pitch is to develop the skill to (very) quickly explain an activity or project to anyone.

A common mistake, particularly for those involved in experimental work, is to enthusiastically start talking about a recent and exciting finding, without sufficiently explaining the business interest. This can be a quick way to lose the attention of the listener. Starting with the bigger picture, and placing a finding in a brief context such as the part of the organization, and the business goals, is likely to establish a better connection. Including a clear indication of the next steps is important, and finishing with a request, be it for comments, advice, contacts or ways of getting support, helps to get a conversation going.

4.5 INNOVATION AND LEARNING

Increasing the ability of an organization to be effective at innovation is helped by learning what works, what does not work, and how improvements can be made. There are three areas in which this can be done: at the level of projects, at the level of the organization, and at the level of the organization in relation to the external world. For those at the start of their careers, the level of projects will have the most relevance, but awareness of the other two levels will help to identify good and bad examples of innovation in practice.

Learning from innovation projects

Many projects are based on proposals, which can require significant effort, for example research proposals. During, or after, execution of the proposal it may become clear that the project is not a success: the technical goals were not achieved, or the business incentive changed. While reporting of the work carried out should be completed, there is often not much interest in comparing the project at the end with the assumptions made at the beginning, as those involved are keen to move on to new topics. Making such a comparison, as a standard practice and across many projects and during a number of years, will help improve the quality of the assumptions made in new proposals. That this can be fruitful is seen in the answers in the case studies (Chapter 5) to the question: "Looking back, what would you have done differently?"

Learning from the innovation mechanisms

At the level of teams, departments, or the whole organization, it is important to review how effective the mechanisms are that have been put in place to stimulate and manage innovation. This covers both the process for managing the flow of ideas, as well as the culture in terms of behavior, communications and learning. While many organizations will have annual reviews of goals and outcome, this appraisal should also include the effectiveness of the innovation mechanism. As discussed in Chapter 2, this includes the number and quality of ideas, and the extent to which they are implemented.

In an article primarily devoted to leadership and innovation (Furr and Dyer, 2014), it was concluded that "when competing on innovation, sustainable advantage comes not from the superiority of any particular invention, but from the superior ability of leaders to foster an organization that can learn from mistakes faster, more efficiently, and more consistently than competitors do."

Interaction of the organization with the external world

In a review of forty years experience with innovations and new technology, and in particular of the role of research and its implications for public policy, the main conclusion was that "innovation is fundamentally about long-term learning, and the main challenge for an innovation-orientated R and D policy is creating a good milieu for such learning, within and between enterprises, research groups and other stakeholders. A long-term learning that requires sustained cooperation, in close contact with real applications" (Elg, 2014). In practice, some degree of learning will inevitably take place when different parties interact. The point made by Elg suggests that the opportunities for joint learning should be explicitly addressed in designing collaborations.

Opportunities for such learning to take place are facilitated when the organizations are relatively close to each other, since face-to-face meetings add to the inevitable electronic contact. Concentration of enterprises in a small geographical area is exemplified by Silicon Valley, home to many major corporations, start-ups and venture capitalists. Many other regions have attempted to duplicate the success of Silicon Valley. As noted in Chapter 1, the success of glass manufacture in the thirteenth century on the island of Murano, in the Venetian lagoon, depended on both competition and collaboration: an early example of an innovation hub involving exchange of knowledge and learning.

5 Case Studies

5.1 INTRODUCTION

The previous chapters have provided a set of concepts for understanding early stage innovation, as well as ways of implementing the ideas. In this chapter we look at a range of case studies. These studies are at the level of projects; in contrast many studies of innovation are at the level of companies and strategy. The case studies presented here have two main purposes. Firstly, they illustrate the concepts discussed in the previous chapters. Secondly, they offer experiences from which other organizations can learn.

The case studies encompass three major companies (Albéa, British Steel, General Electric), one company (Avantium) in the SME category (small and medium-sized enterprises), two examples of the commercialization of academic research (Prof. W.A. Carty and Prof. L.T. Thompson), and two examples of social innovations (Municipality of Amsterdam and Prayas Innovation). These examples are from various countries: France, UK, India, the Netherlands and the USA. In line with the title of the book, the examples cover science, technology and other areas.

Those contributing were asked to provide answers to a set of questions. Two contributors chose a narrative format; others addressed the questions in turn. Editing of the responses has been kept to a minimum consistent with clarity in order to preserve as much as possible "the voice of the innovator." This helps to give a feel for early stage innovation in practice; the styles and content also reflect the different types of organization involved.

The questions were:

The idea

1. How did this idea start? (recognition of an opportunity and/or recognition of a need?)
2. Why was the idea considered a potential innovation? (to whom was it new? how new?)

The process

3. What review process was used to judge progress in developing the idea? Formal stage gates? Other?
4. Was the project handed over at a certain point?
5. If so, how was the handover managed: do you recognize the "Valley of Death," where projects die during a handover? How was this managed?

Resources

6. Were sufficient resources available internally?
7. If not, did you acquire resources or find external resources or partners? How?

Culture and trust

8. Developing an idea frequently involves collaborations which require building of trust, as well as effective communication. Any examples or lessons here?
9. Diversity of inputs in developing an idea is important, but requires an ability to manage the differences. Any examples or lessons here?
10. It is important to balance enthusiasm for the project with being sufficiently critical. How was this achieved?

Follow-up

11. What happened after Proof-of-Concept was established?

Looking back

12. What factors accelerated or hindered development of the idea?
13. If you could do this again, what would you do differently?
14. Any other points you would like to make?

The eight case studies are now given, followed by a discussion of the themes which emerge across them.

5.2 THE CASE STUDIES

Albéa: Highly Reflective Laminate for Cosmetic Tube Packaging

Philip Ashman
Technical Director, Albéa Laminate Tubes

Introduction

The Albéa group offers the world's largest range of beauty and personal care packaging, with leadership positions in plastic and laminate tubes, mascaras, lipsticks and foam pumps. This includes recognized expertise in fragrance and skincare dispensing pumps, fragrance and lotion caps, compacts, lip-gloss, promotional items and accessories.

The group employs 15,000 people, serves large and small customers in 60 countries, has 38 facilities in Europe, Asia and the Americas and global sales of US$1.4 billion (2016).

Albéa has robust capabilities in innovation and design as well as project management, solid operating excellence, world-class manufacturing and a strong commitment to Corporate Social Responsibility.

The idea

How did this idea start? (recognition of an opportunity and/or recognition of a need?)

Albéa's tubes business has a global leadership position in a very competitive market, with innovation being a critical success factor for big and small consumer brands across the world. Differentiation is a key focus for its customers, and "premiumized" (value enhancing) graphics are a way to make their products stand out on the shelves.

A growing demand has been identified for tubes with large printed areas having a multi-colored metallic aspect. These had previously been printed using conventional technology, but at a high manufacturing cost.

This Case Study describes an alternative graphic solution, where the metallic-printing method could be replaced by using a new highly reflective material for the tube itself. Instead of applying metallic print to selective parts of the design, these areas could be printed with a transparent or colored varnish, allowing the reflective properties of the laminate to show through.

The advantage of the alternative approach was the simplification of the metallic printing process. This had previously used "hot-foil" or "cold-foil" technologies, where aluminum metal was transferred from a carrier film to the print surface using hot-melt or UV-cured adhesives.

For tinted-metallic graphics, specialist multi-colored carrier films were available, but the color range was limited and the application technology was not simple to operate at an industrial scale. These factors had an effect of constraining growth for tube products in the very active Beauty Care markets.

The alternative solution was identified from market studies of a range of packaging materials in various global regions and led to the development of a new mirror-effect laminate for the cylindrical body of the tube. The new laminate concept had a highly reflective aluminum metallic layer embedded just below the surface of a laminate material, thereby retaining the heat-sealable properties of the outer and inner layers.

Using this highly reflective plastic laminate, either opaque or color-tinted inks could be applied using conventional flexographic or digital decoration technology, then converted into tubes using existing installed equipment.

This concept also led to a reduction in the supply-chain complexity, a wider range of design options and an unlimited spectrum of colors, all with minimal investment cost.

Why was the idea considered a potential innovation? (to whom was it new? how new?)

The idea of using metalized films as part of laminates for other packaging applications has been widely adopted, particularly for snack foods and confectionary and also for self-adhesive labels. However, these were mostly for thin-film applications, which did not have the specific haptic and heat seal requirements needed for tubes.

For tube applications, some silver laminates were available, but their surface qualities were inferior and their process characteristics were less able to deliver the high quality requirements of the cosmetics market.

By contrast, the new laminate had a mirror-like surface and was specifically designed to include the challenging heat-seal requirements of the tube-making process.

The process

What review process was used to judge progress in developing the idea? Formal stage gates? Other?

The companies' formal stage-gate New Product Development (NPD) process was adopted, through Conception; Concept Justification; Design and Prototyping; Pilot Production; Industrialization; and Production Ramp-up.

As part of this standardized NPD process, key generic actions were defined for each of the project stages to ensure that the inputs from each part of the business were considered, and that any risks identified at the earliest possible stage.

At the completion of each project stage, gate reviews were undertaken to ensure that the key actions for the project stage had been completed.

The project team included a Needs Manager (a Sales and Marketing specialist), a Project Manager, a Product Manager (a technical specialist), a Process manager, a Supply Chain Manager and a Purchasing Manager.

Moreover, the project was selected as a "strategic project" which has a specific governance, consisting of a monthly project review by the Innovation and Development committee.

Was the project handed over at a certain point?

This project involved three handover points.

The laminate was designed internally, then manufactured externally by a supplier. The first handover was the close collaboration with the supplier to ensure that all aspects of the laminate specification and manufacturing requirements had been taken into account.

Once produced, the laminate was delivered to an Albéa manufacturing plant for conversion into unfilled tube packaging. This second handover focused on the process development needed to reach the targeted quality and output.

The final handover point is between Albéa and its customer, once unfilled tubes are delivered to a filling site. At this stage, the customer qualifies the product for product stability and compatibility and checks the filling characteristics when running through high speed packaging lines, all taking into account the changes in the new packaging materials being introduced.

If so, how was the handover managed: do you recognize the "Valley of Death," where projects die during a handover? How was this managed?

The three handover points were recognized as high risk situations in each of the three handover points for the laminate development project. The Failure Mode and Effect Analysis approach (FMEA) was used to identify potential failure scenarios and to assess their impact on quality, supply and financial impact. In this methodology, each failure scenario is given a ranking to determine the severity of the outcome, the likelihood of its occurrence and the probability of detection. Action plans are then created to reduce the overall risk.

FMEA was particularly important for the lamination step, where an alternative technology was adopted to achieve the incorporation of the metallic film layer into the tube laminate. The FMEA helped to identify the priority areas where enhanced levels of control were required. Prior to pilot scale production, close collaboration with the laminate supplier ensured that the additional controls were implemented as standard practices.

Further downstream, for tube conversion and filling, where existing processes were adapted to run the new laminate material, a standardized qualification process was used to verify that acceptable levels of performance had been reached. Alongside this, continued involvement of the Product and Process Managers was essential to make sure all possible adaptations were incorporated into the product or process to overcome any small issues identified during the qualification trials.

In summary, the "Valley of Death" was overcome thanks to:

— a strong project governance

— a good involvement of the manufacturing, sales and marketing teams in the various development phases.

Resources

Were sufficient resources available internally?

Project resources were identified as part of the NPD Project Management system adopted in the early stages of the development, so we were able to predict and budget accordingly.

If not, did you acquire resources or find external resources or partners? How?

Many of the resources employed included those provided by external companies, for example the laminate supplier, process equipment suppliers and external laboratories.

Resourcing within external companies can sometimes be an issue and it was particularly important to have a strong commercial engagement at an early stage in the project to deliver the service levels needed to avoid delays.

Culture and trust

Developing an idea frequently involves collaborations which require building of trust, as well as effective communication. Any examples or lessons here?

The Market Study and Business Plan, which are built into the Project Charter, are key elements to build trust of all stake holders. This requires time spent at the beginning of the project, with involvement of marketing, sales, technical, manufacturing teams and external collaboration. However, this early work saves time for the following development phases, thanks to the engagement of all parties.

Internal and external collaborations were key to the success of the project. Internal collaboration was already well established, as most team members had worked together on other projects. The external collaborations required frequent face-to-face visits and a mutual understanding of business needs. Non-Disclosure Agreements were put in place at the outset of the project to allow a free flow of information amongst the parties.

Diversity of inputs in developing an idea is important, but requires an ability to manage the differences. Any examples or lessons here?

The development had a clear Project Charter document to focus activities on a defined end-point. With that in place, and with a set of deliverables defined in the Stage-Gate NPD process, the team worked in a common direction.

It is important to balance enthusiasm for the project with being sufficiently critical. How was this achieved?

A data-driven approach, with early information regarding financial opportunities was adopted as part of the NPD system. This created a focus on business deliverables and the appropriate success factors from the outset.

Involvement of the top management from the beginning of the project and this engagement created a strong commitment of all participants and helped to maintain the enthusiasm.

Follow-up

What happened after Proof-of-Concept was established?

The phase which follows the Proof-of-Concept is well defined, with the main project management emphasis changing towards implementation and a closer involvement of the manufacturing and supply chain teams.

In this case, Proof-of-Concept was achieved through validation steps built into the NPD process. To ensure the best possible product performance, alternative raw material options were tested to optimize the product resistance and process characteristics of the laminate.

Finally, the product was developed and launched and is now part of a major consumer brand with a global presence.

Looking back

What factors accelerated or hindered development of the idea?

The product launch took longer than expected, mostly resulting from extended approval timelines by customers and planning constraints for development trials using industrial equipment.

Next time, a more aggressive approach to the project planning could be adopted, possibly including the use of pilot scale process equipment recently installed by some of the companies' raw material suppliers.

However, the main factor to accelerate the New Product Development is customer support and commercial success. In this case, the development benefited from a commercial opportunity from a major customer. This gave traction to the industrialization phase, which needs commercially viable quantities to get started.

If you could do this again, what would you do differently?

At the outset, the project would have benefited from an earlier involvement of the Manufacturing and Quality partners which were to become key

Fig. 5.1. Highly reflective laminates developed by Albéa for cosmetic tube packaging.

contributors later on. This would improve the planning ahead of hand-over and get input into risk analysis and process control improvement planning at an earlier stage.

Avantium: Microfluidics in Catalyst Testing

P. Imhof (VP Strategic Account Management) and
R. Moonen (Senior Design Engineer), Avantium Chemicals B.V.

Introduction

Avantium, based in Amsterdam, the Netherlands, was formed in 2000 as a spin-off from the Shell Technology Center in Amsterdam. The initial objective was to accelerate and exploit the application of high-throughput R and D, developed by Shell for catalysis research, across a range of industries.

As of 2017, Avantium is focused on providing advanced catalysis services and systems to companies in the oil, gas, chemical and renewable sectors, as well developing and commercializing new technologies for manufacturing biobased chemicals and plastics.

Innovation at Avantium

Technology and innovation are key business drivers at Avantium. Processes, equipment and methodologies are developed to allow Avantium's businesses to introduce new renewable chemistries processes within Avantium or support Avantium's customers in the exploration and development of new and existing chemical processes.

Avantium uses a stage gate review process to evaluate progress in development of the innovation idea and to decide on follow through at each stage gate and/or milestone. Deliverables for each stage have been described and are monitored during regular New Product Development (NPD) meetings. In these meetings recommendations are made to the management team on whether and how to proceed. Though a formal stage gate process is applied, flexibility is built in to allow maximum efficiency and progress. Decisions on follow up are based on resource availability, potential of innovation, progress and remaining challenges. Only before commercialization a formal milestone check and approval loop on deliverables is executed.

One example has been selected to demonstrate Avantium's approach and the drivers in the innovation process for the different departments.

Use of microfluidics for high quality feed distribution in parallel catalytic testing

Traditionally most of the testing in heterogeneous catalytic processes is executed in fixed bed reactor systems. Within Avantium systems have been developed to accelerate catalysis research by employing a multitude of small scale reactors in parallel, each operated under similar conditions, like feed composition, feed flow, reaction pressure and temperature. With the reactor size applied at Avantium typically only milligram amounts of catalyst are used, while commercial reactors operate at ton scale. It has been demonstrated that by applying the high throughput concept, trends in catalyst performance could be determined, which are characteristic for the behavior at pilot plant and/or commercial scale operations, thus allowing the discovery and optimization of catalytic processes and catalysts in a very efficient way.

In these parallel reactor systems feed distribution was typically achieved by use of multiple feed devices like pumps or by restrictive forces, i.e. capillaries. Though suited for the purpose, use of both multiple pumps and capillaries is labor-intensive and could result in undesired fluctuations in feed distribution.

Avantium has recognized the potential of microfluidics for feed distribution, although several technical challenges needed to be addressed (e.g. the use of high pressure and use of high viscous liquids). The development project included not only mechanical parts, such as the microfluidics and the high precision metal components, but also electronics and software. The integrated system makes use of a control algorithm to ensure exact similar liquid feed flow rates to each of the 16 reactor channels. Any deviation in flow rate will be compensated by the control algorithm.

Overall microfluidics have resulted in a more robust and user-friendly application of feed distribution devices with significantly increased accuracy of the distribution, less maintenance, increased uptime, while increasing the flexibility with respect to feeds and flows.

The project to introduce microfluidics in catalytic testing is typical for Avantium: from concept in prototype in relatively short time championed by the NPD team (6 months), while costs for prototyping and resource were relatively low by using partners in each phase and testing at own facilities and equipment before external commercialization to avoid the "valley of death."

Early involvement of potential customers was crucial to understand the required performance and range of operation. Acceptance was achieved through close involvement of users and other stake holders throughout each phase and testing in the existing equipment and comparing with traditional approaches. This resulted in early recognition of the benefits. Moreover, this could serve as proof to external parties supporting commercial introduction.

After Proof of Concept, the consequences for manufacturing, quality control, retrofits, and new products, IP, and manuals were identified in various teams and implemented in the various processes before official commercial introduction. Effects of introduction of the technology were closely monitored and mitigated, where necessary.

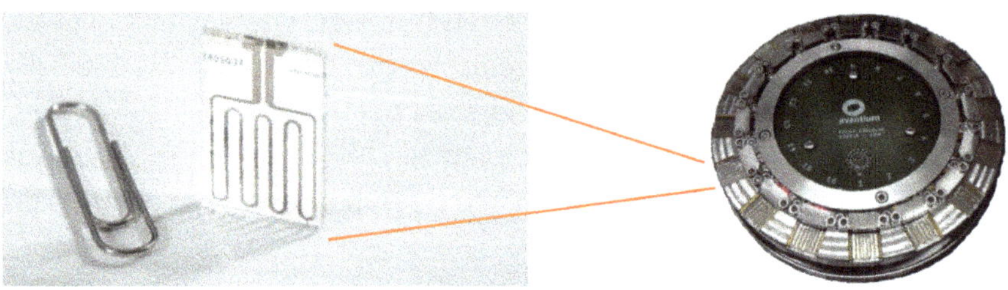

Fig. 5.2. Microfluidic chip developed by Avantium for liquid feed distribution. 16 chips are applied for distribution of liquids to 16 parallel reactors.

British Steel: SilentTrack

David Benton and Dave Farrington (retired),
British Steel, Scunthorpe, UK.

Introduction

SilentTrack® is an innovative noise reduction rail system offered by British Steel. SilentTrack employs patented "tuned dampers" which have been developed in conjunction with Southampton University's Institute of Sound and Vibration Research. The dampers are attached to either side of the rail — absorbing vibrations efficiently and preventing noise from being generated by the rail. Tuned dampers can be fitted to a variety of standard rail sections. By using advanced acoustic modelling techniques, SilentTrack can be optimized to deliver maximum effectiveness on any track configuration, using a variety of installation methods.

SilentTrack has been extensively installed onto tracks both in Europe and worldwide, and more than 175 km of track now benefit from the system, which is a significantly more attractive option than noise containment measures such as barriers that are visually intrusive and very expensive. SilentTrack is recognized by many railway networks as an important method of complying with noise legislation and environmental planning regulations.

The idea

How did this idea start? (recognition of an opportunity and/or recognition of a need?)

The need for railway noise reduction was well established from both environmental concerns and potential legislation, and had received some attention from bodies engaged in railway research. This means that the need was already established, and that it represented an opportunity to apply the know-how available.

Why was the idea considered a potential innovation? (to whom was it new? how new?)

The idea of changing the physical design of the rail system to reduce noise in general had already been investigated, for example using enclosures, or adding damping media. In this case the SilentTrack system was considered innovative (a step forward), because it deliberately targeted the dominant frequencies of noise generating vibration.

The process

What review process was used to judge progress in developing the idea? Formal stage gates? Other?

The development process was considerably assisted by being the subject of a European Commission (EC) Framework project. A very formalized process of results review was followed with a deliberate mid-term step giving approval for continuation.

Was the project handed over at a certain point?

No, the development team remained involved.

If so, how was the handover managed: do you recognize the "Valley of Death," where projects die during a handover? How was this managed?

Not applicable.

Resources

Were sufficient resources available internally?

No.

If not, did you acquire resources or find external resources or partners? How?

The project needed the support of EC research funding, which was obtained through open competition. This included also partners, primarily railway noise specialists from Southampton University, and also rail operators.

Culture and trust

Developing an idea frequently involves collaborations which require building of trust, as well as effective communication. Any examples or lessons here?

Again, the formalized template of the EC framework program forced good communication. There was an existing rapport between the various bodies because of previous work.

Diversity of inputs in developing an idea is important, but requires an ability to manage the differences. Any examples or lessons here?

There were very diverse backgrounds, especially for prototype manufacture. It required a very "hands-on" approach for liaison, but there was nothing particularly novel about this. On the other hand, partners with too similar commercial objectives may have felt constrained from an IP point of view, so that relationships had to be clear and formalized from the start.

It is important to balance enthusiasm for the project with being sufficiently critical. How was this achieved?

End-user involvement from the start was the major contributor to making sure that a realistic approach was taken. In addition there were clear budget constraints.

Follow-up

What happened after Proof-of-Concept was established?

Increasingly realistic tests were carried out to verify the results from the laboratory, culminating in full scale field trials. A huge amount of work was carried out to verify cost-effective production methods. The design for installations on track was subsequently modified substantially to improve speed of installation (critical for railways).

The first prototypes were produced in 2000, followed by on-track trials and in-service monitoring for 2–3 years after this. The first significant commercial installations were in the autumn of 2005.

Looking back

What factors accelerated or hindered development of the idea?

One of the initial fields of study, to carry out measurements of rather esoteric material properties (of a range of different elastomers), was found to be difficult and could not be bypassed.

If you could do this again, what would you do differently?

This was a project which had to combine disparate partners and gain external funding, so it would be difficult to criticise the structure. Perhaps one simplification would be to not run too many optional (noise mitigation) solutions under one umbrella project, however that was built into the bid/funding process so unavoidable in this case.

Any other points you would like to make?

Overall the project was a clear success and partners were happy about their contributions.

GE: Development of a Flexible Array Eddy Current Sensor

Shyamsunder Mandayam, Principal Scientist, GE Global Research, John F. Welch Technology Centre, Bangalore, India.

All engineering components and structures undergo a quality control process where several nondestructive methods are utilized to inspect the component, and qualify it to be free from any flaws which could impair its future performance during service. Among the several NDT (Non-Destructive Testing) methods in use, the Eddy Current Technique (ECT) is a very important nondestructive evaluation and inspection methodology for the inspection of electrically conductive materials, with a capability to detect surface and subsurface defects, carry out material sorting, measure coating thickness, etc. In conventional eddy current testing, alternating current injected into a coil creates an alternating magnetic field. When the coil is placed over an electrically conductive part, eddy currents are generated in the material. Defects in the material perturb the path of the eddy currents causing a change in the impedance of the coil which is measured and correlated.

The technique's versatility and sensitivity makes it suitable for several applications in a variety of industries, including aerospace, power, nuclear, oil and gas, transportation and infrastructure. Traditionally, this inspection can be applied to flat surfaces, rods, wires, tubes and other components. When applied to flat surfaces the sensor/probe size is of the order of a few millimeters.

Sometime in the early to mid-1990s, the need for faster, improved and more reliable inspection of several critical aircraft engine components was emphasized as part of an Engine Structural Integrity Program (ENSIP). Moreover, there was also a drive to move to image-based approaches, which were perceived to be advantageous over the traditional signal-based approach. A typical ECT inspection involves doing a raster scan of the surface with a single surface probe to get an image. However, this is time-consuming and leads to poor productivity. These factors became the key motivation to start exploring newer eddy current sensors and sensor configurations which could overcome the disadvantages of a single probe scan to get images as the inspection output.

The "Array of Sensors" concept was what originated as a potential solution and very soon it became clear that the use of Eddy Current Array Probes (ECAPs) to inspect larger surface areas was the most efficient approach. The simplest implementation of the concept was to electronically drive multiple conventional eddy current coils placed side by side in the same probe assembly. Data acquisition is performed by multiplexing the eddy current coils in a manner to avoid mutual inductance between the individual coils. Creating such a configuration with several rigid probes was not envisaged to be a difficult task but could at best be termed an incremental innovation. However, the team realized that what would be truly disruptive and a breakthrough innovation with real benefit and utility would be when the entire array was on a flexible substrate as it would provide the advantage of conforming to complex and curved surfaces, such as the ones encountered in aircraft engine components. With the benefits of single-pass coverage, and enhanced imaging capabilities, ECAP technology on flexible substrates could provide a remarkably powerful tool and significant time savings during inspections.

Right from the start it was very clear that the road to accomplish this revolutionary innovation was a long one, and needed the active participation and contribution of several domain experts spanning: Physics, Electronics,

Specialized Manufacturing, Signal/Image Processing, Modeling and Simulation, NDT application and shop floor inspectors. Pulling together a multidisciplinary team with such varied skill sets was a challenge which needed patience and perseverance.

The first few versions of the technology were developed based on empirical calculations and the limited fabrication technologies available at that time. Once the initial Proof-of-Concept was established through extensive trials on calibration blocks to demonstrate and assess the performance, the big challenge was to create a roadmap to overcome all the observed deficiencies and with a vision towards its end application in a shop floor/field environment.

Since the development was a collaborative one involving the R&D team and the Engineering team, there were several processes in place including weekly progress reviews, quarterly reviews with senior leadership, and toll gate reviews at major milestones by Chief Engineers. These rigorous processes ensured a steady progress in the right direction with adequate opportunities to make design and other changes at regular intervals to achieve the goals. Since the development was iteration between design, development, fabrication, testing/validation and field trials, the handoff from the R&D team to the Engineering team was not a single step but a continuous process with constant feedback and critical observations.

There were two very important elements whose availability and adoption during the development process gave a tremendous boost to the evolution of the final product. One was the systematic use of "Design for Six Sigma (DFSS)" concepts and tools for all aspects of the development, and the second was the extensive utilization of several well-proven finite element modelling software packages which had become available in the market during the late 1990's. These two factors were instrumental in achieving the outcome: a high performance and near zero defect ECAPs which was very essential for the success of this venture.

Given the fact that the initial developments for this product happened at a time when the company was generally resource-constrained, it was a difficult task to convince the senior management to dedicate the right quality and quantity of resources. Highlighting the potential benefits and the competitive advantage which would result from this innovation once successful was the only way forward and the team was successful in doing that effectively. The resources once allocated had the challenge of accomplishing this innovation in a time-bound manner with a very controlled expenditure that would help gain the confidence of the management.

The major developments in this technology innovation happened at a time when working in global teams was not so prevalent as it is today. Members from different countries with very different backgrounds and cultures were on the same team and the entire experience was new for almost everyone. Building mutual trust and carrying out effective communication was essential to make meaningful progress on the venture. It certainly took a lot of time and effort to accomplish this but once done, it was a wonderful experience to see the team work, mutual understanding and the bonhomie which was created amongst the global teams. While technology was undoubtedly the key to the success of this innovation, the end goal could never have been achieved without giving due attention, importance and credit to the elements of culture, trust, communication and teamwork.

With respect to the time-scales involved, it took us about two years from the first conceptualization of idea to proof-of principle for internal purposes. It took a further two years to be able to modify, improvise and demonstrate a more robust and practical version and the feasibility to a wider audience. Thereafter the developments were continuous and almost every year we had newer and better performing versions, overcoming the gaps observed, and which were field deployable. The developed product was used for the first five years at GE only. There was no commercial introduction to the external world during that period. Later it was offered to GE's partners. The development and the innovation resulted in a portfolio of nearly 10 patents several of which are still active.

The primary commercial impact has been an 8–10x increase in productivity because of faster and more reliable inspection of critical components (it is difficult to put a $$ value on this but the benefits have been accruing for the last 10+ years). The technology continues to be used by GE for several of its engine lines. Interestingly it has not been superseded by any other new technologies.

If we look back briefly, there were several factors in the then prevailing scenario which accelerated the development of the idea. One key factor was the realization and onset of globalization as a theme for growth. Globalization suddenly opened access to a large resource pool of highly qualified technologists around the world. In this specific case study, the opening of an R&D Center by GE at Bangalore, India in early 2000 played a catalytic role. The incubation of a team of scientists with background and experience in the area of nondestructive evaluation and its allied areas including modeling/simulation and signal/image processing was probably one of the many force multipliers which integrated technically and culturally as a homogenous team with the existing teams and expertise at GE Aviation, Cincinnati, USA and GE Global Research, Niskayuna, USA in the development, validation, optimization, maturation and successful field deployment of the ECAP technology.

It is worth mentioning here that the model and framework established for global teaming and working towards a successful innovative product development which evolved in this case study was emulated in a few other cases also with similar outcomes.

Municipality of Amsterdam: Innovative Data-Driven Approaches to Poverty Alleviation

Tamas Erkelens, Program Manager Data Innovation, Municipality of Amsterdam.

Introduction

In recent years the concept of "smart cities" has become important, in which information and communications technologies (ICT) are used to improve the development and management of a wide range of services and assets, such as water, lighting, and waste management, as well as schools, health care, libraries and transportation. Some of the initiatives have a strong technical component, for example installation of energy-efficient lighting; others are more focused on people.

The example discussed here deals with an approach developed in the Municipality of Amsterdam (Gemeente Amsterdam), which combined two ambitions: (i) to increase the effort on people-focused smart city initiatives, and (ii) to alleviate poverty. Within this approach, an innovation program was started and defined as "New methods and technologies to alleviate poverty." Approximately 20% of people in Amsterdam are entitled to at least some poverty-alleviation measures, of which there are no less than about 50 in place.

The idea

Meeting the above ambitions had to take into account two factors. Firstly, it is difficult for the local government to identify those who need, and are entitled to benefits, given restrictions on sharing personal data between different parts of government. Secondly, it must be recognized that some citizens do not know they are entitled to social services at all. Both factors have led to an under-utilization of the benefits available.

The key idea discussed in this contribution was the innovative use of data to enhance poverty reduction. Three activities were defined:

(i) developing a dashboard, which presents information on those entitled to poverty reduction measures, as well as information on those making requests for such benefits; these two groups are not necessarily the same.
(ii) identifying partnerships which could help solving organizational inefficiencies, taking into account legal restrictions relating to privacy.
(iii) establishing an effort on web analytics, to track how the relevant websites are used, in order to improve them and make them more user-friendly.

This data-driven approach to poverty reduction was new within the city of Amsterdam, and in particular was new to the three main targeted users of the innovation team: the central department concerned with poverty alleviation within the city government, the local governments of the seven city boroughs (each of which consists of about 150,000 people) who are responsible for implementing poverty-alleviation measures, and the social workers on the ground within the boroughs who had direct contact with people entitled to benefits.

The innovation program on poverty reduction was led by a program manager directly working for the alderman, and who was separate from the central department concerned with poverty alleviation. The program manager gave an assignment to a project team consisting of two internal data analysts, a business analyst and a project manager, who were new to the domain of poverty reduction.

The process

A project plan with milestones was made, and progress by the innovation team was reviewed every two weeks with the poverty reduction program manager. Due to the fact that the project had not been created with sufficient buy-in at the borough level, handover and implementation proved difficult.

Non-governmental social workers on the ground found the dashboard useful, since it gave new insights into where needs were not being met, and allowed targeted initiatives to be put in place, such as publicity for available measures at specific locations. However, at the level of the councils in the boroughs, the handover was not successful. Training was given in use of the dashboard, but staff in the borough councils did not want to spend time on guiding the implementation with the poverty reduction specialists. Typical responses included "looks nice, but I am too busy" and "it does not fit my job description." A political and budgetary conflict about centralization of the organization was one of the factors complicating the handover to the borough council.

Resources

In this project, budget was not the main limitation. The bigger issue was obtaining accurate and up-to-date information on where there were unmet entitlements to poverty-alleviating measures. It is important to note that this information is not at the level of individuals, since this would contravene privacy laws, but at the level of small geographical areas such as a few streets.

Up-to-date, accurate and secure information was not available within the city administration, and efforts were made to access this information from other organizations. After unsuccessful attempts through the tax authorities, and the social security administration, a breakthrough was made when it was found that the required information was available through the national statistics office. This was not initially known within the project.

Culture and trust

It was recognized at the outset that building trust between the innovation team and the central department responsible for poverty alleviation was crucial. Regular communication was maintained, including review of the project plan and milestones achieved. In retrospect however, too much emphasis was placed on short-term steering of the project, and not enough on rigorously defining the project outcomes, the implementation and the preconditions for handover.

Regarding the balance between being enthusiastic and being critical, it became clear that this was a weak point. The program manager of poverty reduction had expected much more from a big data approach, and the project manager from the innovation team did not challenge the expectations of the program manager sufficiently.

Follow-up

An innovative, data-driven approach to poverty alleviation has been successful through its adoption by social workers on the ground. A second, unexpected, positive outcome was the use of web analytics to monitor web traffic and behaviors, and this has now been adopted by the communications department, and led to improved web sites. A third positive outcome has been the identification of new partners, in particular the national statistics office, as a source of critical information.

Looking back

In addition to the successes noted in the previous section, clear areas for improvement were found which will benefit future innovation efforts within the city administration.

Firstly, more attention should be paid to ensuring alignment between the key players at the outset of the project.

Secondly, the project would have benefited from visible support by senior figures who endorse the goals, support the collaborations involved (internal and external), and who ensure implementation of the outcomes.

Thirdly, more formalized procedures, including a more rigorous approach to project management, would have helped.

Fourthly, more effort should have been paid to balancing enthusiasm for the project with being critical about goals and progress.

Prayas Innovation: Yelo Bag, a Solar-Powered Bag that Converts into a Desk

Manish Mathur, Co-Founder and CEO, Prayas Innovation.

Introduction

Prayas Innovation is a company founded with a vision to address the challenges faced by society in areas of energy, education and health. The company aims to improve the quality of life of people in third world countries thus forming a healthier, safer and more civilized society.

Students in rural areas lack access to basic educational infrastructure such as a desk, a bag, and a light source to study at night. We innovate, design, manufacture and implement a multifunctional and unique solution, a solar-powered bag called Yelo that converts into a desk to lighten millions of lives across the globe.

Manish Mathur completed an MBA in Human Resources, and worked in sales and marketing before leaving to start Prayas Innovation.

The idea

How did this idea start? (recognition of an opportunity and/or recognition of a need?)

The idea came in my mind in 2014, when I visited my school and I found that children were still sitting on the floor and studying. I also studied like that, and it was heart-breaking to see the same conditions after sixteen years. I knew how uncomfortable and painful that was. I started working on a solution the same day, and was determined to make a success of this.

Why was the idea considered a potential innovation? (to whom was it new? how new?)

While designing the bag we never thought of any innovation, our focus was only on solving the problem. During the process of testing and piloting we got huge media coverage and because of that we got so many calls from

Africa, Nepal, Bhutan and neighboring countries. Then we realized that India is not the only country that is suffering from such a basic problem and YELO Bag is a global innovation. Now the innovative YELO Bag is considered a breakthrough-product focused on empowering rural education in India. This was such a basic problem which was visible to everyone for many years but unfortunately no one noticed and took action. Initially this innovation was aimed at rural children in India who do not have access to basic educational infrastructure like a school bag, a writing desk (furniture) and a light to study but now we are also getting huge demand in urban and semi-urban areas.

The process

What review process was used to judge progress in developing the idea? Formal stage gates? Other?

The YELO Bag went through many reviews. It took 40 months to reach the final product. The initial idea was only about a desk, but after doing 3 months research in villages we finally identified 3 basic problems:

1) No waterproof school bag for poor children.
2) No tables and chairs in schools.
3) No electricity at night in their home.

So this led us to the realization that we needed to innovate a product which had to be a light-weight, water-resistant bag that can be converted into a writing-desk and powered by a light source.

The design went through much testing, including production of pilot batches, in order to establish durability and strength. This process almost took 24 months to reach a final product.

Was the project handed over at a certain point?

No, we never handed over to anyone. We worked with a professional bag manufacturer who produced the parts, and then provided jobs to poor people for the assembly.

If so, how was the handover managed: do you recognize the "Valley of Death," where projects die during a handover? How was this managed?

Not applicable

Resources

Were sufficient resources available internally?

Yes, we did not have any external help as one thing was clearly in our mind: we do not come from wealthy family companies, and we have very limited budgets. If we need to solve this problem then it has to be with these very limited resources.

Much of the work has been done by volunteers, but we have hired a few engineers and designers. Volunteers are helping us increase the reach of the product. Everyone is trying to promote the product and generate awareness about it in their own particular areas.

Culture and trust

Developing an idea frequently involves collaborations which require building of trust, as well as effective communication. Any examples or lessons here?

This question hits the bull's-eye in our case. Our whole journey was full of collaborations. As the budget was very low in our case, we could not afford any fixed expenses like salaries, office rent, mentors, consultants etc. So we started working at my friend's backyard. We went to many different colleges to hire technical interns who can work for us free of cost, as they have to go through a 3-month compulsory internship during their degree course. As we were doing social innovation, everyone supported us. Soon we started getting media coverage and people started approaching us and started supporting us. Someone designed a free website for us, someone did a press release for us. This is how the whole journey went.

Diversity of inputs in developing an idea is important, but requires an ability to manage the differences. Any examples or lessons here?

Agreed, but in my case it was absolutely clear in my mind what it was that I was looking for. Yes, of course we faced lots of problems due to change of the team every 3 months. But we managed by dividing every task into 3 months. So every time it was clear to us what we needed to do in the next 3 months and what kind of team we would be requiring after 3, 6 and 9 months.

I personally feel that if you are determined and know what you actually need and how you are going to do it, then multiple inputs can easily be filtered.

It is important to balance enthusiasm for the project with being sufficiently critical. How was this achieved?

Absolutely, it is very important to have an enthusiastic team to achieve any goal. At the start of project we were slightly low as we were starting a project with very limited funds. But every time we recruited college students, we got a superb team. The best part was that everyone had seen this problem in their society, and they feel proud that they are member of the team who is working on the solution, and that they will part of history.

Follow-up

What happened after Proof-of-Concept was established?

During the journey we got great media coverage across the globe which made our company very popular especially in India. We were approached by lots of NGOs, charitable trusts and large companies in relation to their CSR activities. Many colleges and institutions invited us to shares the whole journey.

We also received many awards, for example.

1) Indian Innovation Initiative Award.
2) Best Social Startup Idea Award in ASIA.

3) 50 most impact-full social innovators Globally — World CSR Congress.
4) 30 most impact-full innovators — Forbes.
5) Certificate of Merit — World CSR Congress.
6) Featured in Make in India.

This made us proud and full of confidence, and finally we achieved the goal.

We have now manufactured and distributed more than 10,000 bags across India. We have paid for this. The bags are donated to children who belong to very poor families. We plan to approach large companies, governments and NGO's to get further support.

Looking back

What factors accelerated or hindered development of the idea?

Factors which accelerated the development were:

1) Young team with fresh ideas.
2) We knew what we wanted.
3) Support from people because of social innovation.
4) Support from friends and family.
5) No fixed cost

Factors which hindered the development were:

1) Very low budget for innovation.
2) Change of team in every 3 months.
3) Lack of experience of team and mentors.

If you could do this again, what would you do differently?

Yes, we would do some things differently. We learnt many things in the process. Firstly and most importantly we would keep a few permanent people. Secondly, not to do too much testing of the product because every time it can lead to a new design. Otherwise it was a happy journey.

Any other points you would like to make?

We are currently working on another innovation, which is a Waterless Urinals Cartridge. A single cartridge can save up to 175,000 liters a year.

Fig. 5.3. Prayas Innovation: the Yelo bag.

Prof. W.M. Carty: Selective Batching for Commercial Glass Melting

Bill Carty is on the faculty at Alfred University, New York, USA, and holds the title of John F. McMahon Professor of Ceramic Engineering and is the Chair of the Ceramic Engineering and the Glass Engineering Science programs in the New York State College of Ceramics at Alfred University.

His research focuses on traditional and advanced ceramic processing of colloidal and nanoscale powders, the connection of processing to phase evolution and microstructure, grain boundary tailoring, and physical properties of sintered ceramics. The work of his research group in ceramic processing has resulted in significant improvements in the understanding of ceramic forming operations and defect elimination, microstructure control, and dramatic efficiencies in commercial glass melting. In recent years the research has been extended to address non-aqueous ceramic processing and the development of rules that describe metastable grain boundary chemistry.

Introduction

The glass industry is ancient. The first glasses were formed in Mesopotamia about 3500 BCE. From these humble beginnings, glass production progressed from small crucibles to larger crucible melts, eventually to the development of continuous melters to provide molten glass for production scale hand forming operations. The scale of modern commercial glass, however, is huge, on the order of 200–1000 tons of glass produced per day from a single glass tank. The first continuous glass container process was introduced in England in the late 1880s producing approximately 200 bottles per day — current production levels from a single tank are on the order of two million bottles per day. Float glass (where a ribbon of molten glass floats on a reservoir of molten tin) was introduced commercially in the 1950s (by Pilkington Glass works) and is used for the manufacture of windows, mirrors, and items such as shelves and doors. Both container glass and float glass are typically referred to as SLS glass or soda-lime-silica glass and are similar in composition. For other applications, such as reinforcement fiber glass and flat panel (television) glass, a borosilicate glass is used incorporating boron to facilitate forming and enhance specific glass properties.

The production of glass is essentially the same for all commercial glass production. Raw materials are weighed, mixed, and then introduced to the furnace. The batch size varies, but is usually around 2–4 tons of blended materials. An SLS batch typically consists of soda ash (Na_2CO_3), limestone ($CaCO_3$) and sand (SiO_2) in proportions of approximately 1:1:6 (with the specific ratios determined by the manufacturer and based on a number of variables and desired properties of the glass). Other materials are added to adjust the chemistry of the glass contributing to chemical durability, physical properties and color.

The particle size of the starting raw materials ranges from about 200–500 microns (0.2–0.5 mm in diameter). If the particles are too fine, the turbulent atmosphere in the glass furnace can cause the particles to become airborne and will stick to the furnace walls, etc., causing problems in the furnace operation. On heating, the raw materials react to form the glass, but that process typically involves a preferential reaction between soda ash and limestone forming a low viscosity liquid that separates from the sand. After this reaction (or concurrently), the materials decompose evolving CO_2 gas that eventually escapes the melt via bubble rise. The reaction temperature between soda ash and limestone occurs around 800°C; the temperature of the furnace where the raw batch is introduced is typically around 1300°. The peak temperature in the furnace can be as high as 1550°C.

The batch is added to the tank by either pushing the batch into the furnace with a large plow-shaped pusher or added through a screw (auger) continuous feeder causing the raw material batch to mound up at the front of the tank. The high temperature of the furnace entrance (called the "dog house" in the glass industry) causes the materials to stick together. As the heat penetrates the mound of glass batch reactions between raw materials begin (such as the reaction between soda-ash and limestone, or the melting of boric acid) forming a low viscosity liquid that drains out through the holes between the sand particles. Segregation takes place within minutes of the introduction of the batch into the tank, evidenced by floating islands of partially fused sand. Re-homogenization takes hours after that to complete the reactions and produce a batch-free liquid. It is generally accepted that half of the time in the tank is spent in reaction of batch components and the remaining half in homogenization and bubble removal (known as "fining"). Once the glass is homogeneous and free of bubbles, the molten glass is formed into products.

The idea: selective batching

The idea was that by controlling the reaction paths for batch components used to make commercial glass (flat or float glass, container glass, fiber glass, etc.) the time required to obtain glass from raw materials could be significantly reduced. The approach to controlling reaction pathways was to group problematic raw materials together, in order to preferentially promote specific reactions, and to achieve this by granulation. So the term "Selective Batching" is the idea that the batch components would not change, but that they would be grouped using a granulation process. The granules produced are then mixed with the remaining batch components to form the batch.

Selective batching does two things: (1) dramatically reduces batch-free time; and (2) produces a more homogeneous melt. The batch-free time is relatively easy to characterize as unreacted batch components are easily discernible in the glass melt as a particle inclusion. The homogeneity, however, is more difficult to characterize. The sand is the last component to react and dissolve into the melt. The consequence of this is that there are pockets of liquid that are chemically different, usually composed of a higher silica levels than the surrounding liquid. These pockets have different properties such as refractive index, density, thermal expansion, and viscosity. Large scale segregation, common in conventional melting, is eliminated by selective batching, helping to produce a more homogeneous glass early in the melting process. Another indication of batch segregation, not observed in selective batching, is the formation of intermediate crystalline reaction products. If these precipitates remain undissolved, they are referred to as seeds and stones and a cause for rejection (i.e. these are unwanted defects in the glass product). Research on selective batching demonstrated that it works for all types of commercial glasses, including both SLS and borosilicate glass compositions.

How did this idea start? (recognition of an opportunity and/or recognition of a need?)

Research conducted in a seemingly unrelated field (that of glazes on traditional ceramics, such as porcelain) indicated that raw materials in glazes could react along "preferred reactions pathways" that could be beneficial or detrimental. It was proposed that similar preferred reaction paths in commercial glass production could be either exploited or thwarted to improve the rate of glass batch reactions to form glass. One statement by a scientist from US Borax referring to the production of borosilicate glasses triggered the thought process: "You can look into the furnace and see the boric acid melt and run down the glass batch mound to the top of the melt." The initial idea was that if the boron could be tied up in the early stages of melting, this segregation could be reduced, keeping the boron in the melt, and improving the reaction rate.

Why was the idea considered a potential innovation? (to whom was it new? how new?)

The idea was innovative based on the scientific literature on glass batching and reaction paths that indicated the formation of low viscosity phases would speed reaction rates through an increase in diffusion rates. While it is true that diffusion rates increase with decreasing viscosity, the increase in diffusion distances due to segregation eliminates all benefits afforded by faster diffusion rates. Raw material segregation increases diffusion distances by orders of magnitude (from <1 mm to one meter, a factor of about 1000). The literature, however, appears to ignore this point promoting the formation of low viscosity melts.

Selective batching facilitates the formation of high viscosity liquids, or more correctly avoids the formation of low viscosity melts, by encouraging alternative reactions between the raw materials. By forming high viscosity melts, large scale segregation is avoided and diffusion distances remain similar to the initial granule size — on the order of a millimeter. This keeps raw materials and their reaction products in close proximity promoting faster reaction times and better homogeneity.

The process

What review process was used to judge progress in developing the idea? Formal stage gates? Other?

Funding was obtained in 2004 from the US Department of Energy to test the idea. Laboratory work confirmed the concept was viable, resulting in reductions in batch reaction time of between 40 and 90% (i.e. from 8 hours to 45 minutes) depending on the approach. These trials were conducted initially in laboratory experiments with melt levels from 50 g to 2.5 kg.

The DOE proposal required a project plan, established milestones and an overall review (final report) at the end of the project. The project duration was one-half year and included three meetings: kick off meeting, mid-term meeting and project update, and a final meeting reporting research results. The project was a joint university-industry program with two industrial partners. In spite of significant research results demonstrating that the technique worked and produced significant reductions in batch free time, a second proposal for additional funding to work on other areas of the problem (glass uniformity, fining, other glass chemistries, etc.) was rejected. That work was eventually funded through other sources.

Since the preliminary data was obtained via laboratory-scale crucible melts, there was reluctance to pursue further in the glass industry without demonstration on a large scale. Eventually, with industrial support, pilot plant scale trials were conducted showing exceptional results. These were on a scale of approximately one kilogram per hour of glass production under conditions that would mimic an industrial-scale glass tank. The pilot scale facility held 500 kg of glass from which one kg was pulled, continuously, per hour over the course of several days and different batching and furnace operating conditions. Two pilot scale campaigns were conducted approximately 18 months apart.

Was the project handed over at a certain point?

Scale-up was handed over to the industrial partner at which point we continued to consult.

If so, how was the handover managed: do you recognize the "Valley of Death," where projects die during a handover? How was this managed?

The progress was slow and hampered by a new "need," from the industrial partner for modeling data to facilitate their cost model necessary to justify an industrial trial on 200 ton/day tank. In the initial discussions on the process, the crucible data was deemed insufficient thus requiring the pilot-scale trial. After the pilot-scale trials, we were relegated back to crucible melts to generate the data needed to inform the mathematical models. Difficulties in duplicating some of the results in the crucible melts eventually derailed the project and any subsequent trials. (An ironic twist from our perspective: crucible melt data were not acceptable initially, but in the end, crucible melt problems killed the program.)

We found ourselves in a "Valley of Death" scenario. As we progressed down this path, communication with the industrial partner started to break down as it was evidently concluded that they could not repeatedly get the technology to work. It should be noted that they ignored our recommendation on the granulation route, opting instead for an alternative route that we thought would be more expensive and less reliable. We had conducted extensive trials on granulation technology and eliminated several routes as expensive or problematic. We even had to train "experts" in granulation to get an acceptable material for the pilot-scale trials.

In the end, the program was terminated after the drop in natural gas prices in Europe, making the technology "marginal" from an energy savings perspective. What was difficult to quantify, however, was the potential benefit of higher quality glass and increased production yields predicted by the pilot scale trials.

Resources

Were sufficient resources available internally?

We had sufficient internal resources to conduct the laboratory-scale melting trials. We had no control over the industrial resources necessary to conduct production-scale trials.

If not, did you acquire resources or find external resources or partners? How?

Our external partner, in the end, eventually dropped the program. Our view, however, was that the drop in natural gas prices (cited as the primary driving force for abandoning the trials) was unlikely to be sustained suggesting that eventually the technology would be of significant economic benefit.

Culture and trust

Developing an idea frequently involves collaborations which require building of trust, as well as effective communication. Any examples or lessons here?

We trusted our industrial partner and initially had excellent communication. It seems, however, that in the end, their difficulty in justifying continuing work, without posing the problem to us specifically, ended our collaboration. The first indication that there was a problem was the long delay in responses to correspondence (email). I do not see any solution to this issue.

Diversity of inputs in developing an idea is important, but requires an ability to manage the differences. Any examples or lessons here?

The eventual demise of the project was linked to an internal model developed within the company. We did not have access to the model, nor its assumptions, and so could not address any of the assumptions, issues or concerns brought up. A two-way non-disclosure agreement, in place since the beginning of the collaboration, was not sufficient to allow us access to the model.

(As an aside, we were invited to present our technology to another large company. They, however, refused to provide us a two-way NDA, therefore only protecting them, not us. This would have allowed them to "use" our technology without license potentially forcing us to file a lawsuit to regain control — a tactic they were known for. We declined the invitation.)

It is important to balance enthusiasm for the project with being sufficiently critical. How was this achieved?

Hard to judge. It was difficult to not be enthusiastic for a technology that generates such a significant advantage. We did acknowledge the problem of scale: the smallest commercial glass tank is approximately 200 tons/day and range up to 1000 tons/day. On this scale, a trial would require tens of thousands of tons of raw material with a portion of that being granulated. There were several potential scenarios, of which two were feasible: (1) work directly with a glass manufacturer; or (2) work with a raw material supplier. We are still pursuing the second route.

Follow-up

What happened after Proof-of-Concept was established?

Described above.

Looking back

What factors accelerated or hindered development of the idea?

Economic factors hindered the development eventually, coupled to a predictive mathematical model. Interest was initially fueled by a desire to reduce costs, reduce emissions, and increase throughput for a fixed footprint. In the end, however, it appeared that the energy portion of the equation dominated the discussion.

If you could do this again, what would you do differently?

We would probably start with a raw material supplier for this market. One approach to commercialization would be to supply ready-made granules to the glass industry. For example a granule composed of soda-ash and sand could be used by both the float glass and container glass industry. In similar fashion, a granule composed of boric acid, clay, and sand could be a raw material feedstock for the reinforcement fiber glass manufacturers. In essence, we become a raw material supplier to the industry rather than working with a specific manufacturer to develop an in-house process. We are trying to take this approach for fiber glass production in Asia. In that case we are evaluating setting up a local company to granulate and supply that industry aiming to produce a drop-in batch substitution that incorporates the technology.

Any other points you would like to make?

The scale of the glass industry is staggering. For example, one container manufacturer has 20+ glass tanks in operation in the US. If we assume that (on the low end) the average tank size is 300 tons/day, amounting to roughly 3.3 million tons of glass containers produced annually, the potential savings for the glass container industry could be on the order of \$0.13–0.40/ton, depending on the assumptions used.

The issue, of course, is that with such a scale, the cost for a trial is immense. A normal glass tank holds roughly 2.3–3.0x the pull rate and takes then approximately 3 days to turn over. For a trial that would normally last two weeks, the amount of glass produced, for an average pull rate of 300 tons/day, including the changeover then back to the original operating conditions, would be 6,300 tons of glass (or approximately 140 rail cars in the US). Is the produced product saleable? If not, what are the costs associated with recycling? Is there any liability associated with a new production approach? These questions, and the scale of an industrial trial, show the difficulties of introducing significant innovations in very large scale operations.

Overall we concluded that we had strong evidence for a potentially great technology that was quite informative, but in the end proved not to be industrially manageable. The concept worked remarkably well, but implementation in the commercial glass industry has not been accomplished.

Prof. L.T. Thompson: New Materials for Energy Storage

Levi Thompson is on the faculty at the University of Michigan, USA, and holds the titles of Richard E. Balzhiser Collegiate Professor of Chemical Engineering, Professor of Mechanical Engineering and Director of the Hydrogen Energy Technology Laboratory.* The research in his group focuses on the design, synthesis and characterization of chemicals and nanostructured materials for catalytic and energy storage applications. Of particular interest is the development of structure-function relationships that enable the design of highly efficient materials.

Introduction

In this contribution Professor Thompson describes the first steps in the commercialization of academic research regarding early transition metal carbides and nitrides. This has been carried out through two separate companies, both dealing with energy storage. One company, T/J Technologies, is concerned with new material for batteries; the other, Inmatech, with new materials for supercapacitors.

*As of October 1, 2018, Dean of the University of Delaware College of Engineering and Elizabeth Inez Kelley Professor of Chemical Engineering.

The idea

How did this idea start? (recognition of an opportunity and/or recognition of a need?)

For some years we had been investigating the properties of early transition metal nitrides and carbides including their electronic conductivities and ability to intercalate small atoms. Just before my sabbatical at UC Berkeley I posed that given their high electronic conductivities, facility regarding the incorporation of small ions in their interstitial sites and ability to be produced in high surface area form, early transition metal nitrides and carbides would be effective for energy storage applications. I had the opportunity to evaluate the lithium insertion and double layer storage capacities while at UC Berkeley. While the initial results were modest, I thought that with optimization, the materials could be effective. We have since launched two ventures focused on this class of materials. Much of the early work to develop nitrides and carbides for energy storage was carried out at T/J Technologies, a company that I started with my wife Maria and friend, Dr. Michael Wixom. In 2006, T/J Technologies was acquired by A123 Systems to further commercial materials for lithium ion battery applications. Combined, the organizations represent the largest lithium ion R&D team in North America. A123 Systems continues to be one of the world's leading suppliers of high-power lithium ion batteries for applications including electric vehicles, hybrid electric and plug-in hybrid vehicles, grid stabilization, power tools. A123 Systems had a successful IPO in September, 2009. With the acquisition, I was able to retain the supercapacitor related intellectual property. In 2011, I started Inmatech with Dr. Saemin Choi, a former PhD student, and two former collaborators, Drs. Paul Rasmussen and Stefan Heinemann.

Why was the idea considered a potential innovation? (to whom was it new? how new?)

The lithium ion technology was discovered in the 1980s and since then there has been a flurry of activity to develop better performing materials. By the mid 1990s much of the focus was on oxides and carbons, so the suggestion of nitrides and carbides was new. We filed an application and were awarded our first patent on the technology in 1997 ("High Surface Area Nitride, Carbide and Boride Electrodes and Methods of Fabrication Thereof," U.S. Patent No. 5,680,292, J. Parker, M. Wixom and L. Thompson, 1997 (assigned to T/J Technologies, Inc.). Several other patents followed as the materials were further developed and new applications and materials identified.

The process

What review process was used to judge progress in developing the idea? Formal stage gates? Other?

Much of the early work was supported using federal grants and contracts including Small Business Innovation Research (SBIR) grants. The projects typically including a 6-month feasibility analysis (Phase I), two-year development effort (Phase II) and follow-up commercialization initiative (Phase III). These are essentially stage gates. Within each phase we also established a program plan that included strategic stage gates.

Was the project handed over at a certain point?

Initially I served as CEO of T/J Technologies but just before the acquisition, Maria took over as CEO and I moved to chair the board. She has an MBA and was better equipped for some of the issues that the company was about to face. For Inmatech, we started with Stefan Heinneman as CEO; he is a seasoned entrepreneur with relevant technical experience. I serve as chair of the board but plan to step down within the next year as the company transitions.

If so, how was the handover managed: do you recognize the "Valley of Death," where projects die during a handover? How was this managed?

We were deliberate about the timing of the handover recognizing that a different set of skills was needed moving forward. We did not experience a decline, in fact partnering with A123 accelerated the growth and reach of technology that originated at T/J Techologies. We are pursuing a similar strategy with Inmatech.

Resources

Were sufficient resources available internally?

No. With T/J Technologies, the founders were only able to contribute "sweat equity," i.e. time instead of money. The SBIR grants really made the difference early on. Successes from the grant activities were parlayed into large government contracts, and angel and Series A investments.

If not, did you acquire resources or find external resources or partners? How?

See answer to last question.

Culture and trust

Developing an idea frequently involves collaborations which require building of trust, as well as effective communication. Any examples or lessons here?

Initially the collaborations were primarily between the founders. As we realized our limitations, we sought strategic collaborations with people/entities on the technology and business development fronts. Very recently Inmatech has collaborated with Custom Cell (Germany) to produce prototype supercapacitors and the battery laboratory at the University of Michigan Energy Institute to fabricate advanced electrodes. These were critical to the success of efforts to deliver product to potential joint development partners. We also learned of new challenges and opportunities with regard to our technology.

Diversity of inputs in developing an idea is important, but requires an ability to manage the differences. Any examples or lessons here?

Our core teams have always included an engineer, scientist (typically chemist) and business person. They attend all key meetings including technical meetings. This allows for cross-fertilization and identification of key commercialization issues. Because we started this way, there is an inherent respect for all opinions; even our business development lead making technical suggestions.

It is important to balance enthusiasm for the project with being sufficiently critical. How was this achieved?

Absolutely. We work hard not to become blindly enthusiastic or subject to "group-think." I personally try to be the devil's advocate, looking for ways to get our people to think more broadly and be more critical. I enjoy playing that role.

Follow-up

What happened after Proof-of-Concept was established?

Please see earlier responses.

Looking back

What factors accelerated or hindered development of the idea?

I would say that a lack of experience hindered development of our ideas. We were fortunate to attract funding but frankly a little slow in bringing in the right people. Consequently we have often had to learn on the job. If I were to do anything differently, I would work to optimize the staff's experience even if it meant moving some people out of the company. Clearly the SBIR program helped accelerate development of the concepts and technologies.

If you could do this again, what would you do differently?

I would continually evaluate the staff's experience against the companies' needs and make necessary changes quickly.

Any other points you would like to make?

Participating in entrepreneurial activities has really helped to round out my experience. This experience informs the work that I do as a professor.

5.3 LESSONS FROM THE CASE STUDIES

Here the case studies are reviewed with respect to the concepts presented in preceding chapters. This provides useful illustration of the concepts, and compares different ways by which early stage innovation occurs in practice, from which lessons can be learnt.

In Chapter 1 a definition of innovation was given and the concept of innovation was discussed in terms of the links of innovation to opportunities, needs, problems, adaptations, combinations, and R and D, and in terms of newness. Chapter 2 dealt with the process of early stage innovation, Chapter 3 with open innovation, and Chapter 4 with behaviors, teams, communications and learning.

Meeting the definition of innovation

The definition given in Chapter 1 was "the successful exploitation of new ideas." Of the eight case studies, seven meet this definition. In the eighth case (Prof. Carty) the project failed in the "Valley of Death," and hence does not meet the definition of an innovation. However, current efforts in this project are exploring an alternative route to commercialization for the Proof-of-Concept which was established.

Opportunities, needs, problems, adaptations, combinations, and R & D

The opportunities provided by recognizing the potential of technologies developed for other purposes, and then adapting the technology, are clear in the examples provided by Avantium and the Municipality of Amsterdam.

Recognizing and responding to a need was an important driver in the cases of Albéa, British Steel, GE, the Municipality of Amsterdam, and Prayas Innovation. The innovations developed by British Steel and Prayas Innovation can also be seen as directly addressing a problem: of railway noise, and of very poor learning conditions for children.

In the case of Prayas Innovation, the product Yelo, a bag which can be used as a solar-powered desk, is an example of meeting a latent (unarticulated) need, since the primary customers were not formulating a desire for a Yelo. This is analogous to the latent need addressed by Sony in providing low cost, relatively low quality transistor radios for the teenage market in the 1950's (see Chapter 1).

In the case provided by General Electric the need for better inspection techniques was part of a larger program.

The examples provided by Prof. Carty and Prof. Thompson both rely on scientific insight derived from active research programs, coupled with a recognition of opportunities to apply the insight. In the case of selective batching in glass manufacture (Prof. Carty), the idea was stimulated by making a connection with an adjacent field (ceramic glazes) and this illustrates the benefits that can arise from working in, or being aware of progress in, more than one field at a time. In the case of Prof. Thompson, the opportunity of a sabbatical was used to undertake exploratory research to test an idea.

New to whom? How new?

In some cases the innovation is new with respect to the technology involved, and for the organization offering the innovation, but these aspects will not be of interest to most users: this is likely to be the case for reflective coatings (Albéa), and the use of big data (the Municipality of Amsterdam). In the Avantium example, the application of microfluidics may well be of interest to some users of the equipment; others may be more focused on the benefits this offers.

The innovations provided by Prayas Innovation (Yelo) and British Steel (SilentTrack) have direct impact on the learning environment for children, and on those living near rail tracks. The benefits conferred by the innovation will be appreciated by those affected, who may or may not think about the innovative aspects involved.

In the case of GE, the Flexible Array Eddy Current Sensors had an order of magnitude impact on productivity, and here the newness is likely to be primarily of significance to the GE business.

The innovation projects developed by Prof. Carty and Prof. Thompson are both new with respect to the scientific literature, and in the case of Prof. Thompson the results were also patented, indicating that the criteria for novelty used by the patent granting authority were also met.

In terms of where the innovations fall on a scale between incremental and radical, the examples provided by Prayas Innovation and General Electric, can be regarded as radical innovations. In the case of Prayas Innovation, Yelo is a new and effective solution to a long-standing problem, which has met with recognition and interest from around the world. In the case of GE, work started on an incremental innovation but this led to the idea of a much more radical solution, leading to a patent portfolio, an 8–10x increase in productivity, and to a technology which has not been superseded since its introduction.

It was noted in Chapter 1 that the radical/incremental distinction represents a scale, and not a binary choice. The examples from British Steel, Albéa, Avantium, and the municipality of Amsterdam can all be placed somewhere on a scale between radical and incremental. This is probably also true for the examples from Prof. Carty and Prof. Thompson, although the former example is not yet an innovation; in the case of nitrides for energy storage more information would be required on performance benefits relative to other materials.

None of the case studies reflect purely incremental innovation, but this is unsurprising: authors of case studies will understandably prefer to address bigger steps. However, as pointed out in Chapter 1, the cumulative effect of incremental innovations can be quite significant.

Stage gates and reviews

In all eight case studies rigorous processes were applied to develop the innovation, supporting the idea that innovation is too important to be left to chance, and requires some process to achieve success as efficiently as possible (Chapter 2). In a number of cases the process was imposed by the funding agency: the European Commission (British Steel), Department of Energy (Prof. Carty), and the Small Business Innovation Research (SBIR) program (Prof. Thompson). A formal, stage-gated, new product development process was used by Albéa, and a similar process, modified to increase flexibility, by Avantium. The Municipality of Amsterdam relied on a project plan with milestones; GE used multiple review mechanisms. In developing Yelo, Prayas Innovation depended on many reviews, and on continuous testing and improvements.

Handovers

As discussed in Chapter 2, the handover of early stage projects is often seen as a critical step, frequently referred to as the Valley of Death. An interesting outcome from these case studies is the wide range of approaches used with handover of projects, in most cases successfully.

Albéa identified three handover points, and used a formal process (Failure Mode Effect Analysis) to reduce the risks at handover. Avantium maintained close involvement of users and other stakeholders to ensure acceptance of new technology. For Prayas Innovation, no handover was involved with Yelo, nor was it for British Steel with SilentTrack, where the development team remained involved. For GE, handover was an ongoing, iterative process rather than a one-off event. In the case of Prof. Thompson, the handover was carefully timed, and led to an acceleration of the development. In the example provided by Prof. Carty, the project ended during handover, due to changing perceptions of the industrial partner, and decreasing communication. In the case of the Municipality of Amsterdam, handover to staff at the borough level proved difficult, and this was attributed to insufficient commitment being generated at the borough level at the start of the project.

Use of external resources: open innovation

Seven of the eight case studies report use of external resources in developing the idea. Albéa relied on early engagement with external suppliers, Avantium on close collaborations with partners, British Steel on EC funding, and on specialist knowledge at a university institute as well as collaboration with rail operators. In the case of the Muncipality of Amsterdam, the national statistics office was identified as a key resource during the project. Prayas Innovation (Yelo) worked with an external supplier and brought in volunteers to help develop the idea.

The two examples from academia (Prof. Carty and Prof. Thompson) both involved collaboration with industrial partners. An alternative approach to the commercialization of academic research or insights is through a start-up.

In the case of GE, external partners were not involved, and the result is an example of closed innovation, with many different areas of expertise being available in the same (global) organization. However, the importance of maintaining awareness of external developments is shown by the recognition and adoption of new methods (DFSS, finite element analysis) which became available during the project and which proved to be critical.

Culture and trust: behaviors, teams and communication

The importance of these factors, discussed in Chapter 4, comes across clearly in all the case studies. In two cases (Albéa, British Steel), trust and communication were facilitated by earlier collaborations.

At Albéa the project charter contained elements which addressed building trust amongst all stakeholders; it was noted that this investment of time and effort at the start of the project paid off at later stages. Avantium found that early, and on-going, involvement of potential customers and other stakeholders was crucial. For British Steel, the process imposed by the funding agency (EC) forced good communication between the partners. In the case of GE, it was noted that building trust and effective communication was essential, and required much time and effort. At the Municipality of Amsterdam, it was recognized at the start that building trust between different groups was important, as was maintaining regular communication; however the discussions focused too much on short term adjustments to the project, and not enough on goals and implementation. It was also noted that the balance between being enthusiastic and being critical about goals and progress was too much in favour of the former. At Prayas Innovation, the questions on building trust and communication were said to "hit the bull's eye," as many short term collaborations and volunteers were involved. It was noted that potential problems were avoided by having a very clear vision of the way forward, as well as dividing the tasks up to fit the duration of volunteer availability. In the case of selective batching for glass melting (Prof. Carty), difficulties at the side of the industrial partner in continuing the project were paralleled by decreasing communication, and the end of the collaboration. Prof. Thompson noted the importance of diverse inputs, with a core team composed of an engineer, scientist and business person attending all key meetings. In addition, a deliberate effort was made to avoid becoming too enthusiastic about the project, or become subject to "group-think," aided by Prof. Thompson playing the role of devil's advocate.

Overview of key points

The case studies illustrate the concepts discussed in the preceding chapters, and show a range of approaches and results, which provide opportunities to learn and to improve practices. The main points are summarized in Table 2.

The final question put to the respondents, "any other points you would like to make?" produced some additional supporting material, but in none of the eight cases did it lead to identification of new factors in early stage innovation. This suggests that the set of questions given at the start of the chapter is a useful framework for dealing with early stage innovation projects in terms of:

- Documentation
- Analysis
- Comparison
- Learning

The questions can be applied to completed as well as on-going projects within an organization. In addition, the questions can serve as a basis for discussions with external partners. Developing experience with summarizing projects in this way will help individuals and organizations increase the capability to be innovative.

Table 5.1. Comparison of Key Points from the Case Studies

	The innovation	How did the idea start? (Chapter 1)	Project review process (Chapter 2)	Project handover (Chapter 2)	Use of external resources: open innovation (Chapter 3)	Culture and trust: behaviors, teams and communication (Chapter 4)
Albéa	Highly reflective laminates for cosmetic tube packaging.	Market driven need. Market studies of a range of packaging materials.	Formal New Product Development (NPD) process.	Formal process: Failure Mode and Effect Analysis (FMEA).	Collaboration with suppliers and external laboratories.	Seen as key. Time and effort invested at start saved time later. Existing rapport in internal team.
Avantium	Microfluidics in catalyst testing.	Recognition and adaptation of technology used elsewhere.	Flexible NPD process.	Early involvement of potential customers.	Partners used in each phase.	Close involvement of users and stakeholders.
British Steel	Silent Track: tuned dampers to reduce railway noise.	Well-established need (environmental concerns, potential legislation).	Formal process imposed by funding agency (European Commission).	No handover: development team stayed involved.	European Commission funding. Specialist knowledge at University of Southampton. Rail operators.	Required process forced good communication. Existing rapport between partners.

(Continued)

Table 5.1. (*Continued*)

	The Innovation	How did the idea start? (Chapter 1)	Project review process (Chapter 2)	Project handover (Chapter 2)	Use of external resources: open innovation (Chapter 3)	Culture and trust: behaviors, teams and communication (Chapter 4)
General Electric	Flexible arrays for inspection of complex curved sensors.	Need articulated internally. Work on incremental solution led to a radical innovation.	Rigorous; various types of internal review.	No handover: continuous iteration and feedback.	Essentially closed innovation, but use of new tools which became available externally was important.	Building trust and effective communication in new multidisciplinary, global teams was essential. Required much time and effort to establish, but ultimately effective and rewarding.
Municipality of Amsterdam	Data-driven approach to poverty alleviation.	Derived from strategic ambition.	Project management (but could have been improved).	Difficult handover to one set of users (borough councils).	External partners identified during project.	Insufficient challenge given within the project.
Prayas Innovation	Yelo, a solar-powered bag which converts to a desk.	Determination to solve a long-standing problem.	Many reviews, many designs tested.	No handovers.	Work with external suppliers and with volunteers.	Clear vision and inspirational goal was highly motivating to the project participants.

Prof. Carty	Selective batching for commercial glass melting (Proof-of-Concept established, not yet commercialized).	Scientific insight. Recognition of relevant effects in seemingly unrelated field.	Imposed by funding agency (US Department of Energy).	Handover difficult, project died in the Valley of Death.	Industrial partner.	Started well but communication with partner steadily decreased.
Prof. Thompson	New materials for energy storage.	Scientific insight. Exploratory research during sabbatical.	Imposed by funding agency (US SBIR program).	Handover carefully timed. Led to acceleration of the project.	Industrial partner.	Importance of playing devil's advocate within the team and preventing "group-think."

6 Next Steps

6.1 INTRODUCTION

In this final chapter the steps which can be taken to apply the material presented in previous chapters are discussed, as are some possible future developments in the field of innovation management. Before dealing with these topics it is useful to review some myths about innovation, as well as the pitfalls which can be encountered.

6.2 MYTHS

Here the key points from two articles are summarized, and some additional comments made, as well as links to discussion in earlier chapters.

An article entitled "The five myths of innovation" (Birkinshaw, Bouquet and Barsoux, 2011) discussed myths entitled (i) "The Eureka moment," (ii) "Build it and they will come," (iii) "Open innovation is the future," (iv) "Pay is paramount," and (v) "Bottom up is best."

The myth of the Eureka moment deals with the emphasis on idea generation, and the sudden flash of insight, without paying sufficient attention to the amount of work which is subsequently required. The authors' research supports the widely held view that innovation is 5% inspiration and 95% perspiration. This applies both to efforts within an organization, as well as competitions, or calls for proposals, involving the external world; this was discussed in Chapter 3.

"Build it and they will come" deals with the mixed results obtained from online innovation fora. More generally, this point applies to focusing on opportunities rather than needs, as discussed in Chapter 1. The classic example of the better mousetrap, full of improvements that a customer does not want, illustrates this point.

"Open innovation is the future" notes that open innovation is important, but is not the only way to proceed, and that there is significant effort involved. The article emphasizes the need to combine capabilities in both open and internal innovation; this point was discussed in Chapter 3.

"Pay is paramount." This refers to the discussion of how innovators should be rewarded. One view is that innovators within an organization should be given extra rewards financially, analogous to the extensive rewards which can accrue to employees in successful start-ups. However, this is often not a primary motivator. Many people are motivated by recognition of their efforts, and while this might involve extra financial rewards, visible appreciation, for example in terms of internal awards, publicity and ceremonies can count highly. In addition, many innovators will be spurred on by success, and would like the opportunity to pursue additional innovative ideas.

"Bottom-up is best." Here the point is that the enthusiasm for generating ideas at lower levels in an organization must be balanced by direction from the top. Both are necessary components. In fact it can be an important task for innovation managers to create fora where views and ideas from both the top and the bottom are shared. This can be in the form of electronic communications, or events which senior leaders attend, and provide comments on ideas presented. As discussed in Chapter 1, some ideas will clearly be within the scope of what the organization wishes to achieve, others will clearly be outside this scope, and others will be in a grey area. Establishing a mechanism in which the grey areas can be addressed avoids the frustration of idea-generators who receive no response, and the frustration of senior figures who perceive the ideas as not aligned with carefully formulated strategy.

An article entitled "The seven myths of innovation" (Sawhney and Wolcott, 2004) discussed the following:

1. Myth: you need more new ideas. Reality: You need more homes for ideas.
2. Myth: Innovation is a department. Reality: Innovation is a company-wide competency.

3. Myth: Let people loose to innovate. Reality: Enable people through structure and process.
4. Myth: Innovation is a radical departure from the past. Reality: Innovation often creatively combines pieces from the past.
5. Myth: Mistakes are costly. Reality: Early mistakes are profitable.
6. Myth: Avoid the detours. Reality: Detours may be the destination.
7. Myth: Innovation is about creating new things. Reality: there are many paths to innovation.

These seven points have been addressed at various points in this book but bringing them together forms a useful summary. Myths 1 and 3 are covered by Chapter 2, in how to build the innovation process, and this chapter also includes the importance of weak signals, which is covered by "detours" (Myth 6): "So pay attention to those detours — they may be the faint signals of the next big thing. Myth 2 is covered in Chapter 1, which emphasizes the need within an organization for common concepts and a common language about innovation. Chapter 1 also dealt with combinations of ideas (Myth 4). Myth 5 covers the point made in Chapter 2 that timely termination of an innovation project should not be seen as a failure: termination prevents unnecessary use of resources. As discussed in Chapter 4, it is important to learn from projects which are terminated. Myth 7 deals with the fact that innovation covers more than new products, and this is discussed in Chapter 1.

6.3 PITFALLS

In considering why innovation fails, the following "ten red flags for innovation" have been proposed (Lindergaard, 2013); the order is not significant:

1. The lack of innovation strategy.
2. No common understanding of innovation.
3. Too much focus on internal capabilities.
4. Too much focus on open innovation.
5. Internal silos are too firmly ingrained.
6. Too much focus on ideas and too little focus on people.

7. Lack of a strong networking culture.
8. Innovation efforts focus on technology or products.
9. It's all about the usual suspects.
10. Executives and innovation leaders underestimate the speed of change.

These factors have been discussed in previous chapters. Point 1 is covered in Chapter 2, in terms of agreeing on the scope for innovation. Point 2 is covered in Chapter 1, where it was emphasized that organizations need a common understanding of innovation, and use the same language and definitions. The balance between open and internal innovation (Points 3 and 4) was discussed in Chapter 3, as well in the "myths" covered above. Points 5 and 7 also apply to open innovation, as well as to the issue of managing the internal connections in a large organization, and being able to identify who knows what. This is a particular challenge for new members of staff, and is made worse by reorganizations. Efforts to generate personal pages on websites, or databases of skills and competences sometimes suffer by a lack of commitment to keep them up to date. Point 6, too much focus on ideas rather than people was illustrated (Lindegaard, 2013) with reference to the lack of efforts on matching the right people with the right ideas. Point 6 is also important in relation to open innovation: in considering possible collaborations with external bodies it is important to judge not only the technical value of the collaboration but also the track record of those involved. Point 8 is dealt with in Chapter 1, which emphasizes the breadth of innovation. Point 9 highlights the fact that establishing innovation departments may send a signal that others do not need to be concerned with innovation. On the other hand, separate programs on innovation in many different departments will not be effective unless there is some coordination on division of projects, and on a common language and on measures of success. A premise of this book (Chapter 1) is that a shared awareness of what innovation is, and how it can be achieved, can only help organizations become more effective in being innovative, in terms of outcomes as well as in terms of developing the capability. Finally, response to the speed of change (Point 10) was discussed (Lindegaard, 2013) in relation to the opportunities offered by open innovation. As noted in Chapter 2, identifying and responding to change is an important source of new ideas, as well as being a key part of developing strategy.

6.4 MAKING INNOVATION MORE EFFECTIVE: DISCUSSION POINTS

The material in previous chapters will hopefully give the reader more insight into (early stage) innovation and how to stimulate it. Many of the points covered can also serve as topics for discussion in informal conversations, or in more formal settings, and may lead to involvement of a wider range of participants than would otherwise be the case. Such discussions may be required to reach a particular decision, but can also be valuable in their own right. In this case the conversation is not about establishing right, wrong, or best answers. Instead, the conversations serve to share and sharpen perceptions, and contribute to ongoing learning, ultimately making the organization more effective in innovation.

Some examples of discussion topics are as follows, based on the order of the discussion in the chapters:

- Do we subscribe to the definition of "innovation is the successful exploitation of new ideas" or do we prefer a different definition?
- What examples are there of innovations in our organization?
- How did these innovations start?
- Are these examples best regarded as incremental, radical or disruptive innovations?
- With respect to the process for generating innovations, are there any weak points in our current processes, and do the elements discussed in Chapter 2 offer any ideas for improvement?
- With respect to open innovation, are we making optimal use of external resources to achieve results more quickly, or more cheaply, or results which are better?
- Have we assessed the effort required by our open innovation projects?
- Is our approach to open innovation on a case-by-case basis, or do we have a strategy?

- With respect to teams, are all our innovation teams in the high performing quadrant of Fig. 4.1?
- If not, should the emphasis be on developing more trust, or on more challenge?
- What are the strong and weak points of our internal and external communications on innovation? Are the messages reaching the right people? And in a way which is tailored to the audience or readership?
- When innovation projects are terminated, are key points captured with respect to the assumptions made at the beginning?
- Have we regularly drawn lessons from how our innovation mechanisms work?
- What has been learnt from interactions with the external world? Is there a sustained joint learning as described by Elg (Elg, 2014)?
- Is the format of the case studies in Chapter 5 helpful in documenting our own projects?
- Are there examples of myths, or reasons why innovations fail, which illustrate, or add to, or provide counter examples, to the points given above? Which of the reasons seem the most important? Are individual perceptions of the relative importance of the reasons similar, or different?

This list of discussion topics is by no means exhaustive, and other topics can be found in the preceding chapters. The choice of topics for discussion will change as individual and organizational experience with managing and stimulating innovation develops.

6.5 MAKING INNOVATION MORE EFFECTIVE: CURRENT DEVELOPMENTS

It is very likely that the field of innovation management will continue to evolve, with new examples in practice of effective mechanisms, as well as new insights from academic research. While it is not possible to predict with confidence what these developments might be, the literature reviewed in this book suggests the following topics merit attention; these were discussed in part in earlier chapters; here they are brought together in order to give a clearer picture of possible future developments.

Innovation management processes

The stage-gated funnel described in Chapter 2 is a widely used mechanism to filter and develop innovative ideas, and stage-gates, or review points, are a common feature of new product development and project management in general. As the discussion in Chapter 5 shows, stage-gates, or an equivalent, are used in many of the examples covered. There is increasing interest in applying Agile methodologies to make the traditional stage gates more flexible and adaptive (Chapter 2); it remains to be seen what the impact of this work will be.

A second development of interest concerns Technology Readiness Levels (Chapter 2). Their widespread use in the current European Union Horizon 2020 program will provide information on their effectiveness.

The widely-used models of stage-gates, Technology Readiness Levels, and open innovation which have been discussed in Chapters 2 and 3, are all linear. Although these linear models are helpful in decision-making, it has long been recognized that they are inadequate descriptions of the innovation process (see Chapter 1). Simple linear models do not capture the messy, but realistic, processes shown in Fig. 2.1. More elaborate models have not found wide acceptance (Godin, 2006, 2017).

It thus remains to be seen whether a new model for innovation will emerge which combines linear progression through decision points as well as the multiple interactions and feedback loops characteristic of the innovation process. Two areas of practice with feedback loops are those of beta testing of prototypes and software, in which user feedback is used to improve the original version, and the relatively recent area of translational medicine. Translational medicine has been described as a "rapidly growing discipline in biomedical research that aims to expedite the discovery of new diagnostic tools and treatments by using a multi-disciplinary, highly collaborative approach" (EUPATI, 2017). This reference also emphasizes that translational medicine is a bi-directional concept encompassing both bench-to-bedside factors, as well as bedside-to-bench feedback which is used to improve new treatments.

New digital tools will doubtless continue to stimulate knowledge sharing and collaboration. One example is the rapidly developing area of augmented reality, AR, in which digital data and images are superimposed on the physical world. Organizations such as Amazon, Facebook, General Electric, the Mayo Clinic and the U.S. Navy are implementing AR with significant results (Porter and Heppelmann, 2017). While AR technology will doubtless lead to new applications which are not yet envisaged, it can readily be imagined that AR will help the process of visualizing and collaborating on prototypes which become the basis of innovations.

Open innovation

As discussed in Chapter 3, open innovation has made big steps over the last 15 years, and is increasingly the norm. It has been argued that the term may disappear, simply because it will become the predominant mode of innovation, with closed innovation being relegated to a niche. At the same time, there are still a number of areas where open innovation can be explored further (Chesbrough, Vanhaverbeke and West, 2017; Vanhaverbeke, 2017; Busarovs, 2013), such as further work on open innovation and SMEs (small and medium-sized enterprises), open innovation in relation to developing countries, and open social innovation.

Analysis of innovation management at a project level

Chapter 5 illustrates early stage innovation management at the level of projects, and compares the development of innovations across a variety of topics and organizations. This appears to be a useful approach. With respect to open innovation, analysis at the level of projects has been recommended, with a plea for industries to make their documentation on projects available for researchers aiming to understand how open innovation works in practice (Vanhaverbeke, Du, Leten and Aalders, 2017).

6.6 SOME VIEWS ON THE FUTURE OF INNOVATION

A thought-provoking collection of over 200 essays on the future of innovation has been provided (Stamm and Triflova, 2009), written by authors from different backgrounds in about 60 countries. Each contribution gives a view on the future of innovation. The wide diversity of topics and perspectives involved makes any summary difficult, especially as the points which seem significant will depend on the reader's interests and experience, both of which are likely to change with time. The following examples relate to some of the points discussed in earlier chapters of this book.

Three trends were seen as challenging (Bessant, 2009), and all are still relevant:

(i) Sustainability: this requires both incremental and radical innovations, in products, processes and business models, and covering for example waste, energy, resources and transportation (see Chapter 1).
(ii) Managing the knowledge spaghetti: here the challenge is dealing with vast increase in technological knowledge, together with the global distribution of this knowledge, which leads to the need for effective connections and knowledge brokering. Companies will need to invest effort in finding relevant knowledge, and this point is illustrated with examples in Chapter 3.
(iii) Co-creation with users. This highlights the fact that many users of products are not passive consumers, and that involving them in the front-end of innovation will become increasingly important, one example being the move towards mass customization. User involvement also occurs in the beta-testing of software (Chapter 3), and, as noted above, in translational medicine.

A contribution from the director of the IBM research laboratory in Zurich stated that "the future of innovation is open, multidisciplinary, global and collaborative," and illustrates this statement with examples from some of the many (about 35) collaborations IBM then had within EU framework programs (Kaiserswerth, 2009). While this statement is certainly true for many large companies, small companies may not need to depend on global collaborations.

The need for an "innovation chain master" was covered by a contribution from Philips Applied Technologies (Hissel, 2009). Here it was also noted that innovation in the future will be far more global and based on multi-party collaborations, but this will require management across the chain, to avoid problems from differences in processes, culture, systems, and trust. Such a role might help with the types of problems encountered in handovers, and often described as the Valley of Death (Chapter 2) either within an organization, or in handover to external parties; note, however, the successful project handovers described in Chapter 5.

Turning to issues of trust and communication (Chapter 4), one essay discussed the future of innovation and the disappearing water cooler (Lawrence, 2009). The point was that increasing digital communication and remote working were replacing the traditional gathering of people around a water cooler (or coffee machine). Such informal face-to-face meetings build trust and relationships where new ideas can be tested, cross-fertilized and encouraged, and this is less likely to happen when communication consists primarily of short, brief messages between people who do not often see each other. Many will recognize this at least in part, and so a question for discussion is whether the increasing reliance on digital communication is indeed a limitation as suggested, and if so what actions could be taken.

Communication features in an essay which considered that the future of innovation is selling ideas to skeptical audiences (Harley-Martin, 2009). This contribution notes that many good ideas are neglected not because they are intrinsically poor, but because they are poorly articulated, and that what is required is the building of competences and skills in the articulation, validation and planning associated with the idea. This is illustrated both in Chapter 2, in being prepared for application of stage-gate criteria, and in Chapter 4, which deals with the behaviors and communication skills which support innovation.

6.7 IN SUMMARY

Returning to the main points made in the preface, this book has aimed to provide a concise guide to key concepts in innovation, illustrated with practical examples, and supported by extensive literature references. In addition to illustrating specific points made in the text, the reference list also serves to indicate publications and websites where material relevant to this book can be found.

The book has focused on early stage innovation, on the needs of scientists and technologists, and on those working in organizations, but does not focus exclusively on these areas, so may also be of interest to others. The book aims to help individuals, and their organizations, to become more effective in being innovative, which is reflected both in innovations as an output, as well as building the capability to produce them.

The topic of (early stage) innovation will undoubtedly develop further in coming years, both in terms of understanding and through examples in practice. This book provides a context and foundation for such developments.

References

Allen, R.C. "Collective innovation", 1983. J. Econ. Behaviour and Org., 4, 1.

Amabile, T. 2011. "How to kill creativity", Harvard Business Review, Sept 1998; reprinted in Johnson, S. 2011. "The Innovator's Cookbook", Riverhead Books.

Anthony, S.D. 2012a. "The Little Black Book of Innovation", Harvard Business Review Press.

Anthony, S.D. 2012b. "The new corporate garage", Harvard Business Review, Sept, 42.

Bensen-Amer, R. and Hsieh, T-y. 1997. "Teamwork across time and space". McKinsey Quarterly, 4, 19.

Bessant, J. 2009. "The future of innovation: challenges at the innovation frontier", in von Stamm, B. and Trifilova, A., "The future of innovation", Gower Publishing Ltd.

Bessant, J. and Tidd, J. 2007. "Innovation and Entrepreneurship", Wiley.

Birkinshaw, J., Bouquet, C. and Barsoux, J-L. 2011. "The 5 myths of innovation". MIT Sloan Management Review, 43.

Brunswicker, S. and van der Vrande, V. 2017. "Exploring open innovation in small and medium-sized enterprises", in "New frontiers in open innovation", ed. Chesbrough H., Vanhaverbeke, W. and West, J. (eds), Oxford University Press.

Busarovs, A. 2013. "Open innovation: current trends and future perspectives", Humanities and Social Sciences Latvia, 21, 103.

Business Dictionary. 2017. http://www.businessdictionary.com, accessed 26 Nov 2017.

Cambridge Dictionary. 2017. Organizational Culture, http://dictionary.cambridge.org/dictionary, accessed 26 Nov 2017.

Cellan-Jones, R. 2012. "The future of technology....who knows?", http://www.bbc.com/news/technology-17510101, accessed 26 Nov 2017.

Chesbrough, H. 2003a. "Open innovation: the new imperative for creating and profiting from technology", Harvard Business School Press.

Chesbrough, H. 2003b. "The Era of open innovation", MIT Sloan Management Review, 44, 35.

Chesbrough, H. and Bogers, M. 2017a. in "New frontiers in open innovation", Chesbrough H., Vanhaverbeke, W. and West, J. (eds), Oxford University Press.

Chesbrough, H. 2017b. "The Future of open innovation". Research-Technology Management, 60, 35.

Chesbrough H., Vanhaverbeke, W. and West, J. (eds). 2017c. "New frontiers in open innovation", Oxford University Press.

Christensen, C.M. 1997. "The Innovator's dilemma: when new technologies cause great firms to fail", Harvard Business School Press.

Christensen, C.M. and Raynor, M.E. 2003. "The Innovator's Solution", Harvard Business School Press.

Christensen, C.M., Raynor, M.E. and Anthony, S.D. 2003. "Six keys to creating new-growth businesses", Harvard Management Update, Harvard Business School Publishing, 3.

Christensen, C.M, Raynor, M. and McDonald, R. 2015. "What is disruptive innovation?", Harvard Business Review, 44.

Cooper, R.G. 1990. "Stage-gate systems: a new tool for managing new products", Business Horizons, 44.

Cooper, R.G. 2016a. "Beyond stage-gate: what's next in new-product development", CIMS Newsletter May/June (Center for Innovation Management Studies at North Carolina State University, https://cims.ncsu.edu/); available online.

Cooper, R.G. 2016b. "Next in New-Product Development: Agile-Stage-Gate Hybrids", CIMS Newsletter Nov/Dec (Center for Innovation Management Studies at North Carolina State University, https://cims.ncsu.edu/); available online.

Cooper, R.G. 2017a. "Meeting the challenges of Agile-Stage-Gate-1", CIMS Innovation Management Report, Sept/Oct. (Center for Innovation Management Studies at North Carolina State University, https://cims.ncsu.edu/); available online.

Cooper, R.G. 2017b. "Meeting the challenges of Agile-Stage-Gate-2", CIMS Innovation Management Report, Nov/Dec. (Center for Innovation Management Studies at North Carolina State University, https://cims.ncsu.edu/); available online.

Dahlander, L. and Gann, D.M. 2010. "How open is innovation?", Research Policy, 39, 699.

Davila, T., Epstein, M. and Shelton, R. 2012. "Making innovation work", Pearson Education Inc.

de Jong, M., Marston, N. and Roth, E. 2015. "The eight essentials of innovation", McKinsey Quarterly, April.

Dodd, B. in the foreword to Beacham, J. 2006. "Succeeding though innovation", UK Department of Trade and Industry, http://webarchive.nationalarchives.gov.uk/20070807130014/http://www.dti.gov.uk/innovation/, accessed 26 Nov 2017.

Dodgson, M., Gann, D.M. and Philips, N. (eds). 2015. "The Oxford Handbook of Innovation", Oxford University Press.

Drucker, P.F. 1985. "Innovation and entrepreneurship", Harper and Row.

Drucker, P.F. 2001. "Management challenges for the 21st Century", Harper Business paperback edition.

Drucker, P.F. 2002. "The Discipline of Innovation", Harvard Business Review.

Dugan, R.E. and Gabriel, K.J. 2013. ""Special Forces" Innovation: how DARPA attacks problems", Harvard Business Review, 75.

EARTO, (European Association of Research and Technology Organisations), 2014. "The TRL scale as a research and innovation policy tool, EARTO recommendations".

Economist, 2002. "More Rembrandts in the attic", Jan. 17.

Economist, 2015. "Time to fix patents", Aug. 8th, p.9 and p. 43.

Elg, L. 2014. "Innovations and new technology", https://www.vinnova.se/content assets/e5fe05cb13604be7b221f3ddbecb41c3/va_14_05.pdf

EU Innovation Union. 2017. http://ec.europa.eu/research/innovation-union/index_en.cfm, accessed 26 Nov 2017.

EUAPATI (European Patients' Academy). 2017. www.eupati.eu/non-clinical-studies/translational-medicine, accessed 26 Nov 2017.

European Capital of Innovation award. 2017. http://urbact.eu/european-capital-innovation-award, accessed 24 Nov 2017.

European Innovation Scoreboard. 2017. http://ec.europa.eu/DocsRoom/documents/24829, accessed 26 Nov 2017.

European Research Council. 2017. https://erc.europa.eu/funding/proof-concept, accessed 26 Nov 2017.

Franklin, D. and Andrews, J. (eds). 2012. "Megachange: the world in 2050", Profile Books.

Furr, N. and Dyer, J.H. 2014. "Leading your team into the unknown", Harvard Business Review, Dec, 80.

Furr, N., O'Keefe, K. and Dyer, J.H. 2016. "Managing multiparty innovation", Harvard Business Review, 76.

Galbraith, J.R. 1999. "Designing the innovating organization", Center for Effective Organizations, CEO Publication G 99-7 (366). https://ceo.usc.edu/files/2016/09/1999_07-g99_7-Designing_Innovating_Org.pdf, accessed 26 Nov 2017.

Gaule, A. 2006. "Open innovation in action", H-I Network.
Gertner, J. 2012. "The idea factory: Bell labs and the great age of American innovation", Penguin Press.
Godin, B. 2006. "The linear model of innovation", Science, Technology and Human Values, 31, 639.
Godin, B. 2017. "Models of innovation", MIT press.
Govindarajan, V. 2016. "Planned opportunism: using weak signals to spur innovation", Harvard Business Review, 54.
Grant, A. 2016. "How to build a culture of originality", Harvard Business Review, 86.
Haas, M. and Mortensen, M. 2016. "The secrets of great teamwork", Harvard Business Review, June, 71.
Hall, M. and Johnson, M.E. 2009. "When should a process be art, not science?", Harvard Business Review, March, 58.
Hamel, G. 1999. "Bringing silicon valley inside", Harvard Business Review, 71.
Hamel, G. 2000. "Leading the revolution", Harvard Business School Press.
Harley-Martin, C. 2009. "The future of innovation: selling ideas to sceptical audiences", in von Stamm, B. and Trifilova, A., "The future of innovation", Gower Publishing Ltd.
Harrell, E. 2017. "A brief history of personality tests", Harvard Business Review, March–April, 63.
Héder, M. 2017. "From NASA to EU: the evolution of the TRL scale in public sector innovation", The Innovation Journal: Public Sector Innovation, 22 (2), article 3.
Hender, J. 2003. "Innovation Leadership: roles and imperatives", Grist.
Hiam, A. 2011. "Business innovation for dummies", Wiley.
Hissel, P. 2009. "The future of innovation is innovation chain masters", in "The Future of Innovation", von Stamm, B. and Trifilova, A. (eds), Gower Publishing Ltd.
Huizingh, E.K.R.H. 2011. "Open innovation: state of the art and future perspecitves", Technovation, 31, 2.
Huston, L. and Sakkab., N. 2006. "Connect and develop: inside Proctor and Gamble's new model for innovation", Harvard Business Review, 84, 68.
IBM. 2017. https://www.collaborationjam.com/ibmjam/, accessed 26 Nov 2017.
Imaginatik, see www.imaginatik.com
Imaginatik. 2015. "State of Global Innovation", https://www.imaginatik.com/wp-content/uploads/sites/15/2017/03/state-of-global-innovation-2015-ebook.pdf, accessed 26 Nov 2017.
James, L.K. 1993. "Hermann Staudinger" in "Nobel Laureates in Chemistry, 1901–1992. James, L.K. (ed). American Chemical Society and the Chemical Heritage Foundation.

Jobs, S. 1998, see https://en.wikiquote.org/wiki/Steve_Jobs, accessed 26 Nov 2017.

Johnson, S. 2010. "Where good ideas come from: the seven patterns of innovation", Penguin Books.

Johnson, S. 2011. "The innovator's cookbook", Riverhead Books.

Johnson, S. 2014. "How we got to now: six innovations that made the modern world", Riverhead Books.

Jonker, J. and van Zanden, J.L. 2007. "A history of Royal Dutch Shell, volume 1", Oxford University Press.

Kaiserswerth, M. 2009. "The future of innovation needs to be open and collaborative", in von Stamm, B. and Trifilova, A., "The future of innovation", Gower Publishing Ltd.

Keeley, L., Pikkel, R., Quinn, B. and Walters, H. 2013. "Ten Types of Innovation", Wiley.

Keifer, D.H. 2000. "The establishment of modern polymer science by Wallace H. Carothers", American Chemical Society, (available from: https://www.acs.org/content/acs/en/education/whatischemistry/landmarks/carotherspolymers.html)

Lawrence, J.T. 2009. "The future of innovation and the disappearing water cooler", in von Stamm, B. and Trifilova, A., "The future of innovation", Gower Publishing Ltd.

Leifer, R., McDermott, C.M., Collarelli O'Connor, G., Peters, L.S., Rice, M. and Veryzer, R.W. 2000. "Radical innovation: how mature companies can outsmart upstarts", Harvard Business School Press.

Leonard, D. and Swap, W. 1999. "When sparks fly — igniting creativity in groups", Harvard Business School Press.

Lindegaard, S. 2013. "Ten red flags for innovation: why it fails", www.business-software.com/blog/ten-red-flags-for-innovation-why-it-fails/, July 31st, accessed 26 Nov 2017.

Markham, S.K. and Mugge, P.C. 2015. "Traversing the valley of death: a practical guide for corporate innovation leaders", ISBN 978-0-9909853-1-0.

Mazzucato, M. 2014. "The Entrepreneurial State", Anthem Press.

McAllister, J.F. 1995. "The first nylon plant", American Chemical Society (available from: https://www.acs.org/content/acs/en/education/whatischemistry/landmarks/carotherspolymers.html)

Mortara, L. and Ford, S. 2012. "Technology acquisition: a guided approach to technology acquisition and protection decisions", Institute for Manufacturing, University of Cambridge.

Mostert, N.M. 2015. "Belbin — the way forward for innovation teams", J. Creativity and Business Innovation, 1, 35.

Mulgan G. 2011. "The Process of Social Innovation", in Johnson, S. "The Innovator's Cookbook", Riverhead Books.

Mulgan, G. 2007. "Social innovation: What it is, why it matters and how it can be accelerated", The Young Foundation.

Museum of Failure, 2017. https://www.museumoffailure.se/ accessed 26 Nov 2017.

Noble, C.H., Durmusoglu, S.S. and Griffin, A. 2014. "Open innovation", Wiley.

Obama, B. 2011. https://obamawhitehouse.archives.gov/the-press-office/2011/01/25/remarks-president-state-union-address, accessed 26 Nov 2017.

OECD (Organisation for Economic Co-operation and Development), 2005. "The Measurement of Scientific and Technological Activities, Proposed Guidelines for Collecting and Interpreting Technological Innovation Data", 3rd Edition; also known as the Oslo Manual.

Osterwalder, A. and Pigneur, Y. 2010. "Business model generation", Wiley.

Patanakul, P. and Pinto, J.K. 2014. "Examining the role of government policy on innovation", Journal High Technology Management Research, 25, 97.

Peplow, M. 2017. "Going soft", Chemistry World, 24th Feb.

Phills Jr, J.A., Deiglmeier, K. and Miller, D.T. 2008. "Rediscovering Social Innovation", Stanford Social Innovation Review, Fall 2008.

Pielke Jr, R. 2011. "In Retrospect: *Science — the endless frontier*", Nature, 466, 922.

Pinchot, G. and Pellman, R. 1999. "Intrapreneuring in action: A handbook for business innovation", Berrett-Koehler.

Porter, M.P. and Heppelmann, J.E. 2017. "Why every organization needs an augmented reality strategy", Harvard Business Review, November–December, 47.

Ramirez, R. and Wilkinson, A. 2016. "Strategic reframing: the Oxford Scenario Planning Approach", Oxford University Press.

Ramirez, R., Roodhart, L. and Manders, W. 2011. "How Shell's domains link innovation and strategy", Long Range Planning, 44, 250.

Rigby, D.K., Sutherland, J. and Takeuchi, H. 2016. "Embracing Agile", Harvard Business Review, 40.

Rivette, K.G. and Kline, D. 2000. "Rembrandts in the attic: unlocking the hidden value of patents", Harvard Business School Press.

Rosenman, M.F. 2002. "Serendipity and scientific discovery". In R.D. Norton, (ed.) "Creativity and leadership in the 21st century firm", Emerald Group Publishing.

Rozovsky, J. 2015. "The five keys to a successful Google team", https://rework.withgoogle.com/blog/five-keys-to-a-successful-google-team/, Nov 17th, accessed 26 Nov 2017.

Satell, G. 2016. "The 9 rules of Innovation", Innovation Excellence, http://innovationexcellence.com/blog/2016/05/16/the-9-rules-of-innovation/, accessed 26 Nov 2017.

Sawhney, M. and Wolcott, R.C. 2004. "The seven myths of innovation", Financial Times series, Part Two, on "Mastering Innovation", September 24th.

Sawhney, M., Wolcott, R.C. and Arroniz, I. 2016. "The 12 different ways for companies to innovate", MIT Sloan Management Review, 28.

Schild, P. 2014. http://ec.europa.eu/research/conferences/2013/energy_infoday/pdf/session_3_summary_of_the_calls_open_in_2014_-_philippe_schild.pdf; accessed 26 Nov. 2017

Schwarz, P. 1998. "The Art of the Long View: Planning for the future in an uncertain world", Wiley and Sons.

Schwarz, P. 2011. "Learnings from the long view", Global Business Network.

Shaughnessy, H. 2013. "What makes Samsung such an innovative company?", http://www.forbes.com/sites/haydnshaughnessy/2013/03/07/why-is-samsung-such-an-innovative-company/#2ba80164bb31, accessed 26 Nov 2017.

Shell Pearl GtL. 2017. http://www.shell.com/about-us/major-projects/pearl-gtl/pearl-gtl-an-overview.html, accessed 26 Nov 2017.

Shell, Innovating together. 2017. http://www.shell.com/energy-and-innovation/innovating-together.html, accessed 18 Dec 2017

Sirkin, H.L., Rose, J. and Choraria, R. 2017. "An innovation-led boost for US manufacturing", Boston Consulting Group.

Stadlbauer, M. and Drexler, G. 2014. "De-bottlenecking open innovation: turning patent-based technology network analysis into value", in "Open Innovation", Noble, C.H., Durmusoglu, S.S. and Griffin, A. (eds), Wiley and Sons.

Stiglitz, J.E. 2014. "The Innovation Enigma", www.project-syndicate.org, 9th March.

Stokes, D.E. 1997. "Pasteur's Quadrant: Basic Science and Technological Innovation", Brookings Institution Press.

Technovation. 2017. https://www.journals.elsevier.com/technovation/call-for-papers/special-issue-surviving-the-valley-of-death, accessed 26 Nov 2017.

Thiel, P. (with Masters, B.) 2014. "Zero to One", Virgin Books.

Toegel, G. and Barsoux, J.-L. 2016. "How to preempt team conflict", Harvard Business Review, June, 79.

Trott, P. 2011. "Innovation Management and New Product Development", Pearson.

Tushman, M.L. and Moore, W.L. 1982. "Readings in the management of innovation", Pitman Publishing Inc.

van der Heijden, K. 2005. "Scenarios, the art of strategic conversation", 2nd edition, Wiley.

van Helvoort, T., van Veen, R. and Senden, M. 2014. "Gas to Liquids: Historical development of GTL technology in Shell, Shell Global Solutions International.

van Santen, R., Khoe, D. and Vermeer, B. 2010. "2030: Technology that will change the world", Oxford University Press.

Vanhaverbeke, W. and Chesbrough, H. 2017a. In "New frontiers in open innovation", Chesbrough H., Vanhaverbeke, W. and West, J. (eds), Oxford University Press.

Vanhaverbeke, W., Du, J., Leten, B. and Aalders, F. 2017b. In "New frontiers in open innovation", Chesbrough H., Vanhaverbeke, W. and West, J. (eds), Oxford University Press.

Vanhaverbeke, W. 2017c. "Managing open innovation in SMEs", Cambridge University Press.

Verloop, J. 2007. "Scenarios and innovation" in "Scenarios for Success", Sharpe, B. and van der Heijden, K. (eds), Wiley.

Verloop, J. 2004. "Insight in innovation", Elsevier.

Vlaskovits, P. 2011. https://hbr.org/2011/08/henry-ford-never-said-the-fast, accessed 26 Nov 2017.

von Hippel, E. 1978. "Successful industrial products from customer ideas", J. of Marketing, Jan. 1978, 39 (reprinted in Tushman and Moore, 1982).

von Stamm, B. and Trifilova, A. 2009. "The future of innovation", Gower Publishing Ltd.

Watson, R. 2010. "Future Files: a brief history of the next 50 years", Nicholas Brealey Publishing.

Weddell-Wedellsborg, T. 2017. "Are you solving the right problems?", Harvard Business Review, January–February, 76.

Index

www.ingramcontent.com/pod-product-compliance
Ingram Content Group UK Ltd.
Pitfield, Milton Keynes, MK11 3LW, UK
UKHW020207190926
8624IPUK00034BA/39

9 789813 222021